Operator Theory: Advances and Applications
Vol. 144

Editor:
I. Gohberg

One-dimensional Functional Equations

G. Belitskii
V. Tkachenko

Birkhäuser Verlag
Basel · Boston · Berlin

Authors:

Genrich Belitskii and
Vadim Tkachenko
Department of Mathematics
Ben Gurion University of the Negev
P.O. Box 653
Beer Sheva 84105
Israel

2000 Mathematics Subject Classification 37-xx, 39-xx, 58-xx

A CIP catalogue record for this book is available from the
Library of Congress, Washington D.C., USA

Bibliographic information published by Die Deutsche Bibliothek
Die Deutsche Bibliothek lists this publication in the Deutsche Nationalbibliografie; detailed bibliographic data is available in the Internet at <http://dnb.ddb.de>.

ISBN 3-7643-0084-1 Birkhäuser Verlag, Basel – Boston – Berlin

Member of the BertelsmannSpringer Publishing Group
Printed on acid-free paper produced from chlorine-free pulp. TCF ∞
Cover design: Heinz Hiltbrunner, Basel
Printed in Germany
ISBN 3-7643-0084-1

Contents

Preface

This monograph is devoted to the study of functional equations

$$g(x, \varphi(x), \varphi(F_1(x)), \ldots, \varphi(F_n(x))) = 0, \quad x \in M, \tag{0.1}$$

where M is either the real line $\mathbb{R}$ or the unit circle $\mathbb{T}$, φ is an unknown function, $F_1, \ldots, F_n$ are given mappings of the manifold M into itself and $g : M \times \mathbb{R}^{n+1} \to \mathbb{R}$ is a given mapping.

We assume that the mappings $f, F_1, \ldots, F_n$ are of a class C^k, $k \in \mathbb{N} \cup \{\infty, \aleph\}$. According to the standard definition $C^0(M)$ is the class of continuous functions on M; $C^k(M)$ with $0 < k < \infty$ consists of k-times continuously differentiable functions on M;

$$C^\infty(M) = \bigcap_{1<k<\infty} C^k(M)$$

and, in addition, $\mathbb{C}^\aleph$ is the class of analytic functions on M.

The main question related to equation (0.1) is whether it is solvable in one of the above classes.

Among the most known particular cases of (0.1) there are the classical *equation on implicit functions*

$$g(x, \varphi(x)) = 0, \quad x \in M, \tag{0.2}$$

which does not contain the transformed argument at all, the *Abel equation*

$$\varphi(F(x)) - \varphi(x) = 1, \quad x \in M, \tag{0.3}$$

or the more general *cohomological equation*

$$\varphi(F(x)) - \varphi(x) = \gamma(x), \quad x \in M, \tag{0.4}$$

and the *Schröder equation*

$$\varphi(F(x)) = H(\varphi(x)), \quad x \in M, \tag{0.5}$$

which systematically appears in dynamical problems.

In particular, if $M = \mathbb{R}$ and $F(x) = x+1$, then (0.4) turns out to be a difference equation

$$\varphi(x+1) - \varphi(x) = \gamma(x), \quad x \in \mathbb{R}, \tag{0.6}$$

completely investigated long ago. It is well known that for an arbitrary function γ there exists a solution φ of (0.6) belonging to the same class C^k as γ.

The general solvability problem for equation (0.1) may be divided in two parts: first, to find conditions which guarantee the existence of its local solution in a neighborhood of some point $x_0 \in M$ and, second, granted such local solvability at every point $x \in M$, to find additional conditions for the global solvability on M. Let us illustrate this approach on the equation

$$\varphi(x) = f(x, \varphi(F(x))), \qquad x \in \mathbb{R}, \tag{0.7}$$

in classes C^k, $0 \le k \le \infty$.

If a point $x_0 \in \mathbb{R}$ is such that $F(x_0) \neq x_0$, then it is easy to construct a local C^k-solution in a neighborhood V of x_0. Namely, let $\overline{F(V)} \cap \overline{V} = \emptyset$. Then we can choose an arbitrary C^k-function $\varphi_0(x), x \in F(V)$, define $\varphi_1(x) = f(x, \varphi_0(F(x))), x \in V$, and set

$$\varphi(x) = \begin{cases} \varphi_0(x), & x \in F(V), \\ \varphi_1(x), & x \in V. \end{cases}$$

The function φ may be extended to a C^k-function on a connected neighborhood of the set $\overline{V} \cup \overline{F(V)}$. This yields a local solution since (0.7) is fulfilled for $x \in V$.

If x_0 is a fixed point of F and $y_0 = f(x_0, y_0)$, then various fixed point theorems in functional spaces may be applied to construct a local solution.

If the mapping $F(x)$ has no more than one fixed point and the C^k-mapping $G(x,y) = (F(x), f(x,y))$ is C^k-invertible, the above local solutions can be extended step-by-step via the equation itself as C^k-solutions on the whole axis $\mathbb{R}$. This extension process completely solves the problem of the global solvability modulo local solvability.

The approach described above was applied by many authors to the study of rather general functional equations. Numerous results on solvability of such equations in integrable, smooth, analytic, monotonic, convex functions and on the properties of solutions are represented in monographs [53, 54]. Both monographs contain very detailed lists of references. The current state of affairs is described in the survey [12].

Some simple examples show that a local solvability at every point may not imply global solvability. In particular, this is the case when F in (0.7) has more than one fixed point. A similar effect in the theory of differential equations is known as Stocks phenomenon. This is a situation where an equation has local solutions on overlapping domains covering a neighborhood of a point x_0, but there are obstacles to gluing them to a solution in a whole neighborhood.

Similar obstacles arise for functional equations solvable in a neighborhood of every point $x \in M$ and we describe conditions which permit us to glue such local solutions in a global solution on M.

The material of the book is organized as follows.

Chapter 1 is dedicated to standard questions related to the problem of implicit functions. In addition to the classical theorem on implicit functions we consider relations between formal and local solvability and between local and global solvability.

In Chapter 2 we investigate equations (0.3)–(0.5) closely related to the problem of conjugacy and semi-conjugacy of one-dimensional dynamical systems.

Properties of a mapping which are invariant with respect to transformations of variables determine its dynamical behavior. In turn this behavior determines properties of the related functional equations. We study invariants of mappings and find conditions for two given mappings to be conjugate in the corresponding class of smoothness.

In Section 1, Chapter 2 we consider the fixed-point free diffeomorphisms of the real line and prove that they are conjugate with the shift $x \to x+1$ and that the conjugating transformation may be chosen from the same smoothness class as the diffeomorphisms themselves. While not very complicated for smooth diffeomorphisms, the proof becomes more technically involved for real-analytic diffeomorphisms. For example, one of these proofs appeals to the uniqueness of the smooth structure on the real line, while another is based on the KAM method.

In Section 2, Chapter 2 we investigate equations (0.3) and (0.4) on the real line with arbitrary mappings F, maybe non-invertible. The absence of nonwandering points of F is a necessary condition for the solvability of (0.3) in continuous functions. If F is injective, then this condition is sufficient for the solvability not only of (0.3) but also of (0.4) with arbitrary function $\gamma(x)$. However, if F is not injective, then additional obstacles to the solvability of the Abel equation arise. We introduce a notion of wandering sets for F, and prove that equation (0.4) is solvable in continuous functions for every continuous γ if and only if every compact set is wandering for F.

Section 3, Chapter 2 is devoted to a local classification of mappings in a neighborhood of a fixed point. These investigations were initiated by Poincaré for analytic functions and by Sternberg for smooth functions. It is well known that if a C^k-mapping $F, k \geq 2$, has a hyperbolic fixed point x_0, then it is reducible to a linear form using a local transformation of the same smoothness, including analytic functions. It is known that this statement is not true for class C^1. The results related to the non-hyperbolic case are less known. The first problem arising here is the problem of a formal classification, i.e., the conjugacy in formal power series. We give a complete list of formal normal forms. Further, we describe a new approach to constructing local invariants in a neighborhood of a fixed point of arbitrary type. This approach is based on the conjugacy of F with a standard shift in the left and right semi-neighborhoods of the fixed point. Using a pair of such straightening mappings we construct a local invariant of smooth conjugacy. In this way we state a criterion of C^1-linearization in a neighborhood of a hyperbolic fixed point and construct an infinite family of not smoothly conjugate local diffeomorphisms with a joint fixed hyperbolic point and the same derivative at this point. The same approach permits us to prove that generically the formal C^∞-conjugacy implies the local C^∞-conjugacy. In the analytic case this approach results in the well-known Ecalle–Voronin moduli.

In Section 4, Chapter 2 we study a global classification of one-dimensional diffeomorphisms with isolated fixed points. One of the possible approaches just copies the above scheme using straightening diffeomorphisms on the complement of the set of fixed points and is applicable for fixed points of arbitrary type. If all such points are hyperbolic, then to construct a complete system of invariants a more effective method is used based on the local linearization of diffeomorphisms.

The last section of Chapter 2 is devoted to similar results for one-dimensional flows.

Chapter 3 is dedicated to the study of the equation

$$\varphi(F(x)) = g(x, \varphi(x)) \tag{0.8}$$

whose properties are determined by the dynamical properties of the mapping

$$G(x, y) = (F(x), g(x, y)).$$

In Section 1, Chapter 3 we prove that if $F(x_0) \neq x_0$ and $F'(x) > 0$, then there exists a local transformation

$$\Phi(x, y) = (H(x), T(x, y))$$

which reduces (0.8) to the simplest difference equation

$$\psi(x + 1) = \pm\, \psi(x). \tag{0.9}$$

This allows us not only to prove a local solvability of (0.8) but also to describe the set of all local solutions. A similar problem is studied in a neighborhood of a fixed point, which is traditional in the theory of functional equations started by Poincaré.

In Section 2, Chapter 3 we consider equation (0.8) in the large on $\mathbb{R}$. It turns that if Fix $F = \emptyset$, then a reduction of (0.8) to (0.9) is possible on the entire real line, at least in non-analytic classes.

In Section 3, Chapter 3, linear operators

$$L\varphi(x) = \varphi(F(x)) - b(x)\varphi(x) \tag{0.10}$$

are classified with respect to transformations

$$\Phi(x, y) = (H(x), S(x)y + h(x))$$

which are affine in y and preserve the linearity of (0.8) if the latter was linear. For a classification of linear operators (0.10) we use a "cohomological" approach similar to that used in the problem of conjugacy from Chapter 2. In particular, we state the conditions for solvability of the equation $L\varphi = \gamma$ in cohomological terms applicable both to smooth and analytic functions.

In Chapter 4 we consider equations (0.1) with $n \geq 2$. The properties of these equations are determined by the joint dynamical behavior of the family $\{ F_i \}$. First of all, we introduce the notion of an absorber of the family. The meaning of this notion is clarified by the fact that if $\varphi_0(x)$ is a solution of (0.1) in some neighborhood of an absorber, then under mild conditions of invertibility imposed on $g(x, y_1, \ldots, y_n)$ the function $\varphi_0(x)$ is extendable as a solution of (0.1) for all x. Moreover, every local solution of (0.1) in a neighborhood of the intersection of a system of absorbers is extendable as a global solution. In particular, the solvability holds if the intersection of all absorbers is empty.

The methods of Chapters 3 and 4 are applicable to the equation

$$\varphi(f(x, \varphi(x))) = g(x, \varphi(x))$$

describing the invariant curves in the plane [8, 53], as well as other equations containing superpositions of an unknown function [19].

In Chapter 5 we study the linear operators

$$L\varphi(x) = \sum_{k=1}^{m} a_k(x)\varphi(F_k(x)) \tag{0.11}$$

from the point of view of the general theory of operators in linear topological spaces. We obtain some conditions of surjectivity, of normal solvability and of the Fredholm property for such operators L.

As a matter of fact, many methods used in the present monograph are applicable to and even in some cases first appeared in multi-dimensional situations. Among them there are the gluing of local solutions from Chapters 2 and 3, the KAM method from Chapter 3, the decomposition method from Chapter 4, various theorems on fixed points, and some others.

Let us make several remarks on notation. We use the symbol C^k to denote various classes of smooth or analytic scalar functions defined on some subsets of the real line or circle. If the upper index k is a finite integer, then C^k is the set of all k-times continuously differentiable functions. Class C^∞ is the set of all infinitely differentiable functions. Finally, $C^\aleph$ is the set of all real-analytic functions.

An invertible mapping F will be called the C^k-diffeomorphism if it and its inverse belong to the class C^k. We do not distinguish between C^0-diffeomorphisms and homeomorphisms.

We first became interested in functional equations during our years with the Mathematical Division of the B.Verkin Institute of Low Temperature Physics and Engineering in Kharkov, Ukraine. At the Institute, we enjoyed all advantages of being members of a mathematical community of exceptionally high standards. This interest intensified after we joined the Mathematical Department of the Ben-Gurion University of the Negev, Beer-Sheva, Israel, where many of the results included in the present monograph were obtained with generous grant support from the Israel Science Foundation. Our most sincere acknowledgements go to both of these institutions for their support.

During our student years, we were taught by Professor Yu.I.Lyubich who supervised our work leading to our PhD dissertations. Prof. Lyubich, who organized and chaired many permanent seminars where we presented our first results, eventually became our co-author in many joint publications. We are greatly thankful to him for his involvement in our lives and careers. His influence on us as mathematicians is reflected in this monograph.

Chapter 1

Implicit Functions

An implicit function is defined by an equation

$$g(x, \varphi(x)) = 0. \tag{1.0.1}$$

Here $g(x, y)$ is a given either continuous, or smooth or analytic function of variables x and y, x belongs either to the real axis $\mathbb{R}$ or to the unit circle $\mathbb{T}$, $y \in \mathbb{R}$, and $\varphi(x)$ is an unknown *implicit function.* For existence of $\varphi(x)$ it is necessary, first of all, that for every x_0 there exists at least one solution y_0 of the equation

$$g(x_0, y) = 0. \tag{1.0.2}$$

The following simple example shows that this condition is not sufficient for existence of a continuous implicit function.

Example 1.1. Let $g(x, y) = x^2y - x$. Then for every $x_0 \in \mathbb{R}$, $x_0 \neq 0$, equation (1.0.2) has the unique solution $y = x_0^{-1}$, and arbitrary number $y \in \mathbb{R}$ satisfies it for $x_0 = 0$, but there is not a neighborhood of $x_0 = 0$ where equation (1.0.1) has a continuous solution $\varphi(x)$.

The well-known Theorem on Implicit Functions states that if g is a function from a class C^k, $g(x_0, y_0) = 0$ and $g(x, y)$ is *non-degenerate* at (x_0, y_0), i.e.,

$$\frac{\partial g(x_0, y_0)}{\partial y} \neq 0, \tag{1.0.3}$$

then in some neighborhood of x_0 equation (1.0.1) has a C^k-solution $\varphi(x)$ such that $\varphi(x_0) = y_0$.

Condition (1.0.3) is equivalent to the claim that the plane mapping $(x, y) \to (x, g(x, y))$ is a local diffeomorphism in a neighborhood of the point (x_0, y_0).

The existence of a solution to the linear equation

$$h(x)\varphi(x) + \gamma(x) = 0$$

is equivalent to the divisibility of $\gamma(x)$ by $h(x)$ in the corresponding smoothness class, and (1.0.3) means that $h(x_0) \neq 0$. If the latter condition is not fulfilled, then the ratio $\gamma(x)/h(x)$ may not belong to the same class C^k as $h(x)$ and $\gamma(x)$ and in such circumstances an implicit function exists only if some additional conditions are fulfilled.

1.1 Formal solvability

Let us consider an infinitely-differentiable function $g \in C^\infty$ and let us look for C^∞-solutions of (1.0.1). Assume that

$$g(x_0, y_0) = 0, \quad \frac{\partial g(x_0, y_0)}{\partial y} = 0,$$

and that $\varphi(x)$ is a C^∞-solution of (1.0.1) satisfying $\varphi(x_0) = y_0$. Differentiating (1.0.1) we obtain the additional necessary condition

$$\frac{\partial g(x_0, y_0)}{\partial x} = 0.$$

Differentiating once more we obtain

$$\frac{\partial^2 g(x_0, y_0)}{\partial x^2} + 2\frac{\partial^2 g(x_0, y_0)}{\partial x \partial y} c_1 + \frac{\partial^2 g(x_0, y_0)}{\partial y^2} c_1^2 = 0$$

where $c_1 = \varphi'(x_0)$. In general, (1.0.1) implies relations

$$\frac{d^k}{dx^k} g(x, \varphi(x)) \,|_{(x_0, y_0)} = 0, \quad k = 1, 2, \dots . \tag{1.1.1}$$

If we substitute here $\varphi^{(n)}(x_0) = c_n$, then we arrive at an infinite algebraic system of equations with respect to $c_0, c_1, \dots, c_n, \dots$, and its compatibility is a necessary condition for local solvability of (1.0.1). Let $c_0, c_1, c_2, \dots$ be a solution of the system. According to a well-known theorem by E. Borel [64] there exists a C^∞-function $\varphi(x)$ such that

$$\varphi_0^{(k)}(x_0) = c_k, \quad k = 0, 1, 2, \dots ,$$

and hence the function $g(x, \varphi_0(x))$ satisfies (1.1.1). In other words, $g(x, \varphi_0(x))$ is a *flat function*: its Taylor series at x_0 is trivial. Since, generally speaking, the series

$$\hat{\varphi} = \sum_{k=0}^{\infty} c_k \frac{(x - x_0)^k}{k!}$$

diverges at every point except x_0, we will say that this series is a *formal solution* of (1.0.1) at x_0. It is evident that a function φ_0 with the Taylor series $\hat{\varphi}$ is not unique, but the Taylor series of $g(x, \varphi_0(x))$ depends on $\hat{\varphi}$ only.

1.2 Theorem on local solvability

In this section we consider equation (1.0.1) under the condition that there exists its formal solution $\hat{\varphi}$. Let $g(x,y)$ be a C^∞-function and let $\varphi_0(x)$ be a C^∞-function whose Taylor series coincides with the formal solution $\hat{\varphi}$.

Theorem 1.1. *Let a formal solution* $\hat{\varphi}$ *of* (1.0.1) *at* x_0 *be such that the Taylor series of the function* $\theta(x) = g'_y(x, \varphi_0(x))$ *at* x_0 *is not trivial. Then there exists a* C^∞*-function* $\varphi(x)$ *whose Taylor series at* x_0 *coincides with* $\hat{\varphi}$ *and which satisfies* (1.0.1) *in a neighborhood of* x_0.

Proof. Let us look for a solution of (1.0.1) in the form $\varphi = \varphi_0 + \psi$. It is more convenient to consider instead of (1.0.1) the equation

$$g_1(x, \psi(x)) = -g(x, \varphi_0) \tag{1.2.1}$$

with the function

$$g_1(x,y) = g(x, \varphi_0(x) + y) - g(x, \varphi_0(x)),$$

which may be written in the form

$$\theta(x)\psi(x) + h(x, \psi(x)) = -g(x, \varphi_0(x)),$$

where, as above, $\theta(x) = g'_y(x, \varphi_0(x))$, and

$$h(x,y) = g(x, \varphi_0(x) + y) - g(x, \varphi_0(x)) - \theta(x)y.$$

It is easy to see that

$$h(x,y) = y^2 h_1(x,y)$$

where $h_1(x,y)$ belongs to C^∞. According to the conditions of Theorem 1.1 the function $\theta(x)$ is not flat at x_0 and hence

$$\theta(x) = (x - x_0)^k \theta_1(x)$$

with some entire $k \geq 0$ and a C^∞-function θ_1, $\theta_1(x_0) \neq 0$. If we set $\psi(x) = (x - x_0)^k \psi_1(x)$, then we obtain the equation

$$(x - x_0)^{2k}\theta_1(x)\psi_1(x) + (x - x_0)^{2k}\psi_1^2(x) h_1(x, (x - x_0)^k \psi_1(x))$$
$$= -g(x, \varphi_0(x))$$

with respect to a function $\psi_1 \in C^\infty$. Since the function $-g(x, \varphi_0(x))$ is flat at x_0, we have

$$\psi_1(x) + \psi_1^2(x) h_2(x, (x - x_0)^k \psi_1(x)) = \gamma(x) \tag{1.2.2}$$

with C^∞-functions

$$\gamma(x) = -\frac{g(x, \varphi_0(x))}{(x - x_0)^{2k}\theta_1(x)}, \quad h_2(x,y) = \frac{h_1(x,y)}{\theta_1(x)}.$$

Since the function $y + y^2h_2(x,y) - \gamma(x)$ satisfies condition (1.0.3) at the point $(x_0, 0)$, we conclude that (1.2.2) has a C^∞-solution ψ_1. Differentiating (1.2.2) at x_0 we find that all derivatives $\psi_1^{(k)}(x_0)$ are equal to zero, and hence $\varphi = \varphi_0 + \psi$ is a C^∞-solution with the same Taylor series as φ_0. □

If the function θ is flat at x_0, then the statement of Theorem 1.1 is not true. Indeed, the equation

$$\tau(x)y - \sqrt{\tau(x)} = 0, \quad \tau(x) = \begin{cases} \exp(-x^{-2}) & x \neq 0, \\ 0 & x = 0, \end{cases} \tag{1.2.3}$$

is of the form (1.0.2), has the unique solution $y_0 = (\tau(x_0))^{-1/2}$ at every point $x_0 \neq 0$, and every number y_0 satisfies it at $x_0 = 0$. The corresponding equation (1.0.1) is formally solvable at $x_0 = 0$: every formal series $\hat{\varphi}$ is its formal solution. Nevertheless (1.2.3) has no local continuous solution in a neighborhood of $x_0 = 0$.

1.3 Transformations of equations

If we substitute

$$\varphi(x) = H(x, \psi(x)) \tag{1.3.1}$$

in (1.0.1), we obtain the equation

$$\tilde{g}(x, \psi(x)) = 0 \tag{1.3.2}$$

where

$$\tilde{g}(x,y) = g(x, H(x,y)).$$

If the plane mapping $(x,y) \to (x, H(x,y))$ is invertible, then we have some equivalence relation between equations of the form (1.0.1) or between functions $g(x,y)$. For example, let condition (1.0.3) hold and hence the mapping $(x,y) \to (x, g(x,y))$ is a local diffeomorphism in a neighborhood of (x_0, y_0). According to the Theorem on Implicit Functions there exists the inverse mapping $(x,y) \to (x, H(x,y))$ such that

$$g(x, H(x,y)) = y.$$

In other words, $\tilde{g}(x,y) = y$, and (1.3.2) takes on the simplest form

$$\psi(x) = 0. \tag{1.3.3}$$

Therefore the function $\varphi(x) = H(x,0)$ is the unique solution of (1.0.1).

The situation becomes more complicated if $g(x,y)$ degenerates at (x_0, y_0). In this case, of course, it is not possible to reduce (1.0.1) to the form (1.3.3), and the problem arises of simplifying (1.0.1) by using transformations (1.3.1). Such reduction is equivalent to a classification of singularities and constructing normal forms with respect to a transformation of variables. Here we give a simple example of such a classification.

Example 1.2. Let a C^∞-function $g = g(y)$ in (1.0.1) depend on the variable y only. Condition (1.0.3) means

$$g'(y_0) \neq 0,$$

and if it fails, then y_0 is called a *critical point* of the function g. Let

$$g'(y_0) = \cdots = g^{(k-1)}(y_0) = 0, \quad g^{(k)}(y_0) \neq 0, \quad k > 1.$$

It is well known that, depending on whether k is either an even or an odd integer, y_0 either is or is not an extremal point of g. In a neighborhood of y_0 the function g behaves similarly to the simple function

$$h(y) = g(y_0) + \frac{g^{(k)}(y_0)}{k!}(y - y_0)^k. \tag{1.3.4}$$

More exactly, there exists a locally invertible C^∞-transformation $\Phi(y) = y + \varphi(y)$ which reduces g to the normal form (1.3.4):

$$g(\Phi(y)) = g(y_0) + c(y - y_0)^k, \quad c = \frac{g^{(k)}(y_0)}{k!}.$$

To find Φ, let us introduce the inverse map $\Psi(y) = y + \psi(y)$, $\Phi(\Psi(y)) = y$. Since

$$g(y) = g(y_0) + c(y - y_0)^k + f(y - y_0), \quad f(t) = o(t^k),$$

we arrive at the equation

$$g(y_0) + c(y - y_0)^k + f(y - y_0) = g(y_0) + c(y - y_0 + \psi(y))^k$$

and obtain its solution

$$\psi(y) = (y - y_0)\left(\left(1 + \frac{f(y - y_0)}{c(y - y_0)^k}\right)^{1/k} - 1\right).$$

The function ψ belongs to class C^∞ and $\Psi(y) = y + \psi(y)$ defines the transformation we looked for.

Example 1.3. Let

$$g(x,y) = y^p + y^{p+1}v(x,y) + \gamma(x), \quad x_0 = y_0 = 0; \; v(0,0) = 0; \; v, \gamma \in C^k.$$

Let us show that the function g may be reduced to the form $g_1(x,y) = y^p + \gamma(x)$. To this end let

$$H(x,y) = y + y^2\theta(x,y)$$

with some function $\theta \in C^k$. Then the equation

$$g(x, H(x,y)) = y^p + \gamma(x)$$

leads us to the relation

$$\theta + \tilde{v}(x, y, \theta) = 0 \tag{1.3.5}$$

where

$$\tilde{v}(x, y, \theta) = -\frac{(1 - y(1 + y\theta)^{p+1} v(x, y + y^2\theta))^{1/p} - 1}{y}.$$

Since $\tilde{v}$ belongs to class C^k and $\tilde{v}(0,0,0) = 0$, there exists a solution $\theta \in C^k$ of (1.3.5). Substitution (1.3.1) reduces (1.0.1) to the form

$$(\psi(x))^p + \gamma(x) = 0$$

whose study is not difficult.

1.4 Global solvability

Following our general concept we consider here the following problem. Let (1.0.1) have a solution (either continuous or smooth or analytic) in a neighborhood of every point $x_0 \in M$. Is its solution defined everywhere on M?

The simplest examples show that without additional restrictions the answer is negative.

Example 1.4. Let us consider the nonlinear equation

$$(x^2 y - 1)((x + 1)^2 y + 1) = 0.$$

The function $y = x^{-2}$ is its unique continuous solution on the set $U_1 = \mathbb{R} \setminus \{0\}$, and $y = -(x + 1)^{-2}$ is such a solution on $U_2 = \mathbb{R} \setminus \{-1\}$. Nevertheless the equation has no solutions continuous on the real line.

It is obvious that if a local solution is unique in a neighborhood of every point, then the set of all local solutions is glued in a global solution on M. The uniqueness of a local solution is guaranteed by condition (1.0.3).

Theorem 1.2. *Let there exist for every $x_0 \in M$ a root $y = y_0$ of* (1.0.2), *and let $g'_y(x, y) \neq 0$ for points $x \in M, y \in \mathbb{R}$. Then there exists the unique C^k-solution of* (1.0.1) *defined on M.*

Proof. According to assumptions of the theorem a local C^k-solution of (1.0.1) exists in a neighborhood of every point $x \in M$. It is sufficient to prove that such a solution is unique. But if y_1 and y_2 are solutions, then

$$0 = g(x, y_1(x)) - g(x, y_2(x)) = g'_y(x, \theta(x))(y_1(x) - y_2(x))$$

and $y_1 = y_2$ follows. □

Let us note that neither condition of Theorem 1.2 follows from the other.

Example 1.5. Let $g(x,y) = e^y - \gamma(x)$. Then $g_y'(x,y) = e^y \neq 0$, but (1.0.1) has a solution if and only if $\gamma(x) > 0,\ x \in \mathbb{R}$.

Of course, the solvability of (1.0.2) for every $x_0 \in M$ does not imply (1.0.3). Moreover, for global solvability it is not sufficient to request that there exists a unique root y_0 for every $x_0 \in M$.

Example 1.6. Let $g(x,y) = (xy - 1)(x^2 + y^2)$. Equation (1.0.2) has the unique root $y_0 = x_0^{-1}$ for every $x_0 \neq 0$ and $y_0 = 0$ for $x_0 = 0$, but it has no continuous solutions defined everywhere on $\mathbb{R}$.

The situation is essentially simpler for linear equations.

Theorem 1.3. *Let the equation*

$$f(x)\varphi(x) = \gamma(x) \tag{1.4.1}$$

have a C^k-solution in a neighborhood of every point $x_0 \in M$. Then it has a C^k-solution φ defined everywhere on M.

Proof. The statement is evident in the analytic situation, since the zero set of $f(x)$ is discrete and all its zeros are of finite order. The local solvability in a neighborhood of a point x_0 where f has a zero of order k means that x_0 is a zero of γ of order not less than k. If this is the case for every x_0, then $\gamma(x)/f(x)$ is defined and analytic everywhere on M.

In a non-analytic situation an arbitrary closed set may be the zero set of $f(x)$. To prove Theorem 1.3 in this case we use a *partition of identity* which is not applicable for analytic functions. Namely, let φ_α be a C^k-solution of (1.3.5) in a domain $U_\alpha \subset M$, and let $\{\tau_\alpha\}$ be a partition of identity subordinated to the covering $\{U_\alpha\}$. This means that the non-negative function $\tau_\alpha \in C^\infty$ vanishes outside a closed subset $V_\alpha \subset U_\alpha$ for every α and

$$\sum_\alpha \tau_\alpha(x) \equiv 1, \quad x \in M.$$

Then the function

$$\varphi(x) = \sum_x \tau_\alpha(x)\varphi_\alpha(x)$$

belongs to class C^k and is a solution of (1.4.1) everywhere on M. □

1.5 Comments and references

The local theorem on implicit functions is classical and has a multi-dimensional version for mappings $g : \mathbb{R}^n \times \mathbb{R}^p \to \mathbb{R}^m$ with some modification of condition (1.0.3). Namely, it is sufficient to request that the *Jacobi matrix*

$$\frac{\partial g(x_0, y_0)}{\partial y}$$

be of rank m.

Theorem 1.1 is valid in a multi-dimensional case if, in addition to its conditions, an inequality

$$\|\theta(x)\| \geq c\|x - x_0\|^k$$

holds with some $c > 0$, $k > 0$, for all x from a neighborhood of x_0. The latter estimate is usually called the *Lojasiewicz inequality*. According to the *Lojasiewicz Theorem* [56] this condition is fulfilled for real-analytic functions if x_0 is an isolated zero of $\theta(x)$. In a general case an inequality

$$\|\theta(x)\| \geq C(\mathrm{dist}(x, N))^k$$

holds where N is the zero set of θ. In an analytic situation the theorem on the local solvability follows from the formal solvability without additional restrictions on dimension and behavior of formal solution (the *Artin Theorem* [10]).

Theorems 1.2 and 1.3 also have multi-dimensional versions.

Chapter 2

Classification of One-dimensional Mappings

In this chapter we study a problem of classification of smooth mappings of the real line $\mathbb{R}$ and of the standard circle $\mathbb{T}$. Such classification is a central problem in the theory of dynamical systems on manifolds, and it is closely connected with functional equations of the form

$$g(x, \varphi(x), \varphi(F(x))) = 0, \quad x \in M, \tag{2.0.1}$$

where $F : M \to M$ is either a continuous, or C^k-smooth or analytic mapping.

According to the well-known definition, two mappings $F, G : M \to M$ are called *C^k-conjugate* (or *conjugate in a class C^k*) if there exists a C^k-diffeomorphism $\Phi : M \to M$ such that the *Schröder equation*

$$\Phi F(x) = G\Phi(x), \quad x \in M, \tag{2.0.2}$$

holds. We will say that F and G are *topologically conjugate* if $k = 0$, that they are *smoothly conjugate* if $1 \leq k \leq \infty$, and that they are *analytically conjugate* if Φ is analytic.

Further, the mapping $G^n(x) = \underbrace{G(G \ldots G(x))}_{n \text{ times}}$ is called the *n-th iteration* of G. Since $G^n = \Phi F^n \Phi^{-1}$ for conjugate mappings, a conjugacy does not change a behavior of iterations.

Given a point $x_0 \in M$, the set $\{F^n(x_0)\}_{n=0}^{\infty}$ is called the *orbit* or *trajectory* of x_0 with respect to the mapping F.

A point $x_0 \in M$ is said to be *periodic* of a *period* p if $F^p(x_0) = x_0$. A fixed point of F is its periodic point of period 1. By Per F and Fix F we denote the sets of all periodic and fixed points of F, respectively.

It is easy to see that if a continuous mapping of the real line has at least one periodic point, then it necessarily has a fixed point. The famous theorem by Scharkovskii [70] states that if a continuous mapping of the real line has a periodic point of a period p, then it necessarily has periodic points of every period $q \prec p$, where $\prec$ is a special ordering in the set of integers. The number 1 is the least and the number 3 is the most with respect to the ordering, and if a mapping has a periodic point of order 3, then it has periodic points of every integer period.

A mapping of the standard circle $\mathbb{T}$ into itself may have periodic points of an arbitrary period p, and no periodic points of any other period. An evident example is given by the rotation $\alpha \to \alpha + 2\pi p^{-1}$ (mod 2π) with period p.

The mappings $x \to \Phi(x)$ and $\varphi(x) \to \varphi(\Phi(x))$ transform (2.0.1) into equation

$$\tilde{g}(x, \psi(x), \psi(\tilde{F}(x)) = 0$$

where

$$\tilde{g}(x, y, z) = g(\Phi(x), y, z), \quad \tilde{F}(x) = \Phi^{-1}(F(\Phi(x))), \quad \psi(x) = \varphi(\Phi(x)).$$

Transformations of such a type are systematically used in the theory of functional equations, which gives an additional motivation to study, along with a mapping F, the class of all conjugate mappings $\Phi F \Phi^{-1}$.

Let $\tilde{F} = \Phi^{-1} F \Phi$ with a diffeomorphism Φ inverting orientation on M and let $\Phi_0 : M \to M$ be another diffeomorphism with the same property. Then $\tilde{F} = \Psi^{-1}(\Phi_0^{-1} F \Phi_0)\Psi$ where $\Psi = \Phi_0^{-1}\Phi$ is a diffeomorphism preserving an orientation. For this reason, if otherwise not stated, we consider a classification of mappings with respect to orientation-preserving diffeomorphisms.

Our main objective in the present chapter is to describe properties invariant with respect to the conjugacy.

The most simple property of a mapping $F : M \to M$ which does not change with transition to a conjugate mapping is the monotonicity. In other words, if F and $\tilde{F}$ are conjugate, at least topologically, and one of them is a monotonic function, then the other is monotonic as well and the type of monotonicity (i.e., increasing or decreasing) is the same.

The number of preimages also is an invariant with respect to conjugacy. Namely, if $\tilde{F} = \Phi^{-1} F \Phi$ and S is a subset, then

$$\Phi \tilde{F}^{-1}(S) = F^{-1}\Phi(S),$$

which means that preimages $\tilde{F}^{-1}(S)$ and $F^{-1}(\Phi(S))$ are homeomorphic. In particular, this implies that if, for some $x \in M$, one of the inequalities

$$\#\{\tilde{F}^{-1}(x)\} > \sup_{t \in M}\{\# F^{-1}(t)\}; \quad \#\{\tilde{F}^{-1}(x)\} < \# \inf_{t \in M}\{F^{-1}(t)\},$$

is fulfilled, then F and $\tilde{F}$ are not conjugate.

If F is a homeomorphism, then $\#\{F^{-1}(x)\} = 1$ for every $x \in M$. If F is strictly monotonic, then $\#\{F^{-1}(x)\} \leq 1$, $x \in M$.

The *identical* mapping $F(x) = x$, $x \in M$, is conjugate with itself only. Another simple mapping of the real line is a *shift* $F(x) = x + c$, $c \neq 0$. The homothety transformation $\Phi(x) = cx$ conjugates it with the *standard shift* $x \to x + 1$.

A quadratic mapping $F(x) = ax^2 + bx + c$, $a \neq 0$, maps the real line on a semi-axis, and all points of its image, except for one, have two preimages. The linear transformation $\Phi(x) = a^{-1}x - (2a)^{-1}b$ reduces F to the form $\tilde{F}(x) = \Phi^{-1}F\Phi(x) = x^2 + d$, with $d = b/2 - b^2/4 + ac$. If $d \leq 1/4$, then $x_{\pm} = 1/2 \pm \sqrt{1/4 - d}$ are fixed points of $\tilde{F}(x)$. Properties of trajectories of quadratic mappings with $d \leq 1/4$ are very complicated and are a subject of intensive and deep study (cf., [57, 58]). In contrast with this situation all quadratic mappings with $d > 1/4$ are conjugate, at least topologically.

For the proof let us construct a homeomorphism conjugating $\tilde{F}$ with the mapping $G(x) = x^2 + 1$. To this end let Φ_0 be an increasing function on $[0, d]$ such that $\Phi_0(0) = 0$ and $\Phi_0(d) = 1$. Since the intervals $I_n = [\tilde{F}^n(0), \tilde{F}^{n+1}(0)]$, $n \geq 0$, cover the semi-axis $[0, +\infty)$ and do not overlap, the relations

$$\Phi_1(x) = G^n \Phi_0 \tilde{F}^{-n}(x), \quad x \in [\tilde{F}^n(0), \tilde{F}^{n+1}(0)], \tag{2.0.3}$$

define a monotonic and continuous function on $[0, \infty]$. Setting

$$\Phi(x) = \begin{cases} \Phi_1(x), & x \geq 0, \\ -(\Phi_1(x^2 + d) - 1)^{1/2}, & x \leq 0, \end{cases}$$

we obtain a homeomorphism of the real line with $\Phi(0) = 0$. If now $x \geq 0$, then $x \in I_n$ for some $n \geq 0$, $\tilde{F}(x) \in I_{n+1}$ and $\Phi\tilde{F}(x) = G^{n+1}\Phi_0\tilde{F}^{-n}(x) = G\Phi(x)$. On the other hand, for $x \leq 0$ we have $\Phi(x) \leq 0$ and $G\Phi(x) = \Phi^2(x) + 1 = \Phi_1(x^2 + d) = \Phi\tilde{F}(x)$. Hence $G = \Phi\tilde{F}\Phi^{-1}$.

Polynomials of odd degree $2n + 1 \geq 3$ necessarily have fixed points. As to polynomials of degree $2n \geq 4$, they may be non-conjugate even in the absence of fixed points. An explicit example of such mappings is given by the functions

$$F(x) = x^4 + 1, \quad \tilde{F}(x) = x^4 - 7x^2 + 6x + 25.$$

It is easy to see that both mappings have no fixed points, but nevertheless they are not conjugate. Indeed, $\#\{F^{-1}(x)\} \leq 2$ for all $x \in \mathbb{R}$, while $\tilde{F}^{-1}(25) = \{0, 1, 2, -3\}$.

Difficulties encountered in a study of classification problems for smooth mappings are connected with existence of a variety of invariant properties which may be different from those discussed above. There are not many classes of mappings for which a satisfactory classification with respect to the conjugacy is known. One of them is formed by invertible mappings without fixed points on the real line and another by diffeomorphisms with isolated fixed points. They are studied in the present chapter.

2.1 Wandering and non-wandering subsets

Given a mapping $F : M \to M$, a point $x_0 \in M$ is called *non-wandering* for F if for every neighborhood U of x_0 there exists a sequence $n_k \to +\infty$ such that $F^{n_k}(U) \bigcap U \neq \emptyset$. More generally, a subset $K \subset M$, is called *non-wandering* for F if for every neighborhood U of K there exists a sequence $n_k \to +\infty$ such that $F^{n_k}(U) \bigcap U \neq \emptyset$.

It is evident that all periodic points are non-wandering, but as a rotation of $\mathbb{T}$ by an irrational angle shows, there exist mappings without periodic points and with a non-void set of non-wandering points.

Because of compactness, each continuous mapping of the standard circle in itself has non-wandering points. On the other hand, the shift $x \to x + 1$ is the simplest example of a mapping of the real line into itself without non-wandering points. We will show (Theorem 2.1 below) that every diffeomorphism $F : \mathbb{R} \to \mathbb{R}$ without fixed points is conjugate with the shift.

From the point of view of dynamical behavior, all monotonic mappings of the real line have a simple structure. Namely, if $F : \mathbb{R} \to \mathbb{R}$ preserves the orientation, i.e., $F(x)$ is an increasing function, then all its non-wandering points are fixed. If F is decreasing, then it has a unique fixed point, F^2 is increasing and its non-wandering points are fixed. In any case all non-wandering points of a monotonic mapping of the real line are periodic of a period not exceeding 2, while non-monotonic mappings may have non-wandering sets of a more complicated nature.

We will define now a notion which, in some sense, is opposite to the non-wandericity and is useful in a study of functional equations.

Definition 2.1. Given a mapping $F : M \to M$, an open subset $U \subset M$ is called *wandering* for F if there exists a number $N > 0$ such that $\overline{F^n(U)} \cap \overline{F^k(U)} = \emptyset$ for $|n - k| > N$.

A set S is called *wandering* if it has a wandering neighborhood.

Every fixed-point free mapping on the real line has no non-wandering points. Moreover, such a mapping has no non-wandering compact subsets. Indeed, if to be definite we assume $F(x) > x$, then $F^n(x)$ monotonically tends to $+\infty$ as $n \to \infty$, and therefore for every bounded set U there exists a number N such that $F^n(U) \bigcap U = \emptyset$ for all $n \geq N$. If, in addition, F is strictly monotonic, then every compact set is wandering in the sense of Definition 2.1. This follows from the equality

$$F^p(U) \bigcap F^q(U) = F^q(F^{p-q}(U) \bigcap U),$$

which holds for every bounded subset $U \subset \mathbb{R}$, and $p \geq q$.

The following examples show that for non-monotonic F the absence of non-wandering compact subsets does not imply that all compact sets are wandering.

Example 2.1. Let

$$F(x) = x + 2 + \cos 2\pi x, \quad x \in \mathbb{R}.$$

Since Fix $F = \emptyset$, F has no non-wandering compact subsets. On the other hand, the interval $I = [2^{-1}, 1]$ is not wandering. Indeed,

$$F^n(I) \supset \left[n + \frac{1}{2}, 3n + 1\right], \quad n = 0, 1, 2, \dots .$$

Hence

$$F^n(I) \bigcap F^{2n}(I) \supset \left[2n + \frac{1}{2}, 3n + 1\right] \neq \emptyset$$

while $2n - n = n \to \infty$.

Example 2.2. Let us set

$$F(x) = x + 1 + (x \sin \pi x^2)^2, \quad x \in \mathbb{R}. \tag{2.1.4}$$

Since $F(x) > x$, the mapping F has no non-wandering compact subsets. Let us show that any interval containing the point $x_0 = 0$ is not wandering for F, which means that this point is not wandering for F. Indeed, if $0 < \xi < 1$, then

$$F^{k-1}([0, \xi]) = [k - 1, k - 1 + \theta_{k-1}]$$

where $\theta_{k-1} > \xi$ for $k > 1$. Let us choose k such that $k - 1 > 4^{-1}\xi^{-1}$ and set

$$z_k = \sqrt{(k - 1)^2 + \frac{1}{2}}.$$

Then $z_k \in F^{k-1}([0, \xi])$ and $F(z_k) \in F^k([0, \xi])$. Since

$$F(z_k) = (k - 1)^2 + \sqrt{(k - 1)^2 + \frac{1}{2}} + \frac{3}{2},$$

we obtain

$$F^k([0, \xi]) \supset \left[k, (k - 1)^2 + \sqrt{(k-1)^2 + \frac{1}{2}} + \frac{3}{2}\right].$$

If now $p_n = n$ and q_n is the entire part of $n^{3/2}$, then $q_n - p_n \to \infty$ as $n \to \infty$, and for all sufficiently large n the inequalities

$$p_n < q_n < (n-1)^2 + \sqrt{(n-1)^2 + \frac{1}{2}} + \frac{3}{2}$$

hold. Therefore

$$F^{p_n}([0, \xi]) \bigcap F^{q_n}([0, \xi]) \neq \emptyset.$$

2.2 Mappings with wandering compact sets

As we have already seen, all compact sets are wandering for strictly monotonic mappings without fixed points. Moreover, this statement holds if either $F(x) > x$ and $F(x)$ is monotonic for $x > N$ or $F(x) < x$ and $F(x)$ is monotonic for $x < -N$ with sufficiently large N. In particular, all compact sets are wandering for every fixed-point free polynomial.

2.2.1 Strictly monotonic mappings without fixed points

Every fixed-point free strictly monotonic mapping F is necessarily increasing and the following alternative holds:

1. F is a homeomorphism of the real line, i.e., it is both injective and surjective.

2. One of the limits $\lim\limits_{x \to \pm\infty} F(x)$ is finite while the other is not.

The simplest example of the former pattern is supplied by the standard shift, while the functions $F(x) = e^x$ and $F(x) = -e^{-x}$ are of the latter pattern. Note that $\Phi(x) = -x$ conjugates them.

Theorem 2.1. *Strictly monotonic continuous mappings F and G without fixed points are topologically conjugate if and only if they are of the same type* 1 or 2. *A conjugating diffeomorphism preserves (or inverts) the orientation of $\mathbb{R}$ if and only if $(F(x) - x)(G(x) - x) > 0$ (or $(F(x) - x)(G(x) - x) < 0$, respectively).*

Proof. Without loss of generality we assume that $F(+\infty) = G(+\infty) = +\infty$. Let $F(-\infty) = c_1$, $G(-\infty) = c_2$. Since F and G are of the same type, then c_1 and c_2 are both either finite or infinite. Assume that $c_1 \neq c_2$, and hence both numbers are finite. Denote by $\Phi(x)$ an orientation-preserving diffeomorphism satisfying $\Phi(c_1) = c_2$. Then F and $\Phi^{-1}G\Phi$ are of the same type and with the same values at $\pm\infty$. Now it is sufficient to consider the case $F(x) > x$, $G(x) > x$ and $F(\pm\infty) = G(\pm\infty)$; the remaining cases may be reduced to it using the auxiliary conjugacy by $\Phi(x) = -x$.

If $F(-\infty) = G(-\infty) = c > -\infty$, then we set $x_0 = c$ and obtain $F(x) > x_0$, $G(x) > x_0$ for all x, while for $c = -\infty$ we choose and fix an arbitrary x_0.

Since $F(x_0) > x_0$, there exists a neighborhood V of x_0, a neighborhood U of the interval $[x_0, F(x_0)]$ and a strictly increasing continuous function Φ_0 on U such that the identities

$$\Phi_0(x) = x, \quad \Phi_0(F(x)) = G(x), \quad x \in V, \tag{2.2.1}$$

implying

$$\Phi_0(F(x)) = G(\Phi_0(x)), \quad x \in V. \tag{2.2.2}$$

If $c = -\infty$, then the relations

$$\Phi(x) = G^n(\Phi_0(F^{-n}(x)), \quad x \in F^n(U),\ n = 0, \pm 1, \pm 2, \ldots \tag{2.2.3}$$

define a homeomorphism of the real line conjugating G and F.

If $x_0 = c > -\infty$, then these relations define a monotonic function Φ on the semi-axis $[x_0, +\infty)$. To extend this function to the entire real axis we set

$$\Phi(x) = G^{-1}(\Phi_0(F(x))), \quad x \in (-\infty, x_0),$$

which is permissible since $F((-\infty, x_0]) = (x_0, F(x_0)]$ and therefore $\Phi_0(F(x)) \in (x_0, G(x_0)] \subset \operatorname{Im} G$ for $x \in (-\infty, x_0)$. The resulting mapping Φ is a homeomorphism conjugating F and G and satisfying $\Phi(\pm\infty) = \pm\infty$. □

The smooth and analytic conjugacy of monotonic mappings of a corresponding smoothness may fail due to the presence of critical points. Indeed, if x_0 is a critical point for F , then $\Phi(x_0)$ is critical for $G = \Phi F \Phi^{-1}$. The number of critical points and their multiplicities are invariant with respect to a smooth conjugacy.

Theorem 2.2. *Let F and G be C^k-monotonic mappings without fixed and critical points. If they are of the same type 1 or 2, then they are C^k-conjugate. A conjugating diffeomorphism preserves (or inverts) the orientation of $\mathbb{R}$ if and only if $(F(x) - x)(G(x) - x) > 0$ (or $(F(x) - x) - x) < 0$, respectively).*

Proof for non-analytic smooth classes is the same as of Theorem 2.1. In addition to the definition of $\Phi_0(x)$ we require $\Phi_0(x)$ to be of the same smoothness class C^k as $F(x)$ and $G(x)$ and to have no critical points.

The definition of $\Phi_0(x)$ based on (2.2.1) cannot be used for analytic functions. Its existence in the analytic situation is a direct consequence of Theorem 3.1 which will be proved in Section 3.1.1. □

If we choose $G(x) = x + 1$ or $G(x) = e^x$, then we arrive at the following propositions.

Corollary 2.1. *Every continuous strictly monotonic mapping F of the real line without fixed points is topologically conjugate either with the standard shift $G(x) = x + 1$ or with the exponential function $G(x) = e^x$ depending on whether it is surjective or not. A conjugating homeomorphism preserves (or inverts) the orientation of $\mathbb{R}$ if and only if $F(x) - x > 0$ (or $F(x) - x < 0$, respectively).*

Corollary 2.2. *Every C^k-mapping F of the real line without fixed and critical points is C^k-conjugate either with the standard shift, or with the exponential function depending on whether it is surjective or not. A conjugating homeomorphism preserves (or inverts) the orientation of $\mathbb{R}$ if and only if $F(x) - x > 0$ (or $F(x) - x < 0$, respectively).*

2.2.2 The Abel and cohomological equations

If the Schröder equation (2.0.2) has a C^k-solution $\Phi(x)$, maybe not invertible, then it is said that F is *semi-conjugate* with G in class C^k. The semi-conjugacy is not an equivalence relation, since in particular it is not symmetric. Indeed, a constant mapping $\Phi(x) \equiv const$ semi-conjugates every mapping F with the identical mapping $G(x) \equiv x$. On the other hand, the identical mapping $F(x) \equiv x$ is semi-conjugate with a mapping $G(x)$ if and only if $\mathrm{Fix}\, G \neq \emptyset$, and if the latter condition is fulfilled, every constant mapping $\Phi(x) = x_0$, $x_0 \in \mathrm{Fix}\, G$, is a semi-conjugating mapping.

The equation of semi-conjugacy

$$\Phi(F(x)) = \Phi(x) + 1 \tag{2.2.4}$$

of F with the standard shift is called the *Abel equation.*

Proposition 2.1. *If the Abel equation has a continuous solution, then all compact subsets are wandering for F.*

Proof. Indeed, if Φ is a solution of the Abel equation, then

$$\Phi(F^p(x)) - \Phi(x) = p, \quad p \in \mathbb{N},\ x \in \mathbb{R}. \tag{2.2.5}$$

According to Definition 2.1 for every compact subset K which is not wandering there exist two converging sequences $\{x_n\}, \{y_n\} \subset K$ and two sequences of integers $\{p_n\}$, $\{q_n\}$, $p_n - q_n \to \infty$, such that $F^{p_n}(x_n) = F^{q_n}(y_n)$. Let $x_n \to z$, $y_n \to u$. It follows from (2.2.5) that $\Phi(x_n) - \Phi(y_n) = q_n - p_n$ which contradicts the continuity (and even the local boundedness) of Φ on $[z, u]$. □

The mappings F from Examples 2.1 and 2.2, Section 2.1 are oscillating as $x \to \pm\infty$. As a result there exist finite closed intervals which are not wandering for F and hence the corresponding Abel equations have no continuous solutions. On the other hand, we will show now that if F is fixed-point free and "monotonic at infinity" then there is no obstacle to the solvability not only of (2.2.4) but also of a general cohomological equation

$$\varphi(F(x)) - \varphi(x) = \gamma(x). \tag{2.2.6}$$

Theorem 2.3. *The cohomological equation with a continuous mapping $F : \mathbb{R} \to \mathbb{R}$ has a continuous solution for every continuous function $\gamma(x)$ if and only if the following conditions are satisfied simultaneously:*

i) F has no fixed points;

ii) There exists $c \in \mathbb{R}$ such that $F(x)$ is strictly increasing either on $[c, +\infty)$ if $F(x) > x$, or on $(-\infty, c]$ if $F(x) < x$.

Proof. NECESSITY. Condition *i*) is evident. Moreover, according to Proposition 2.1 for every compact interval K there exists a number $\nu = \nu(K)$ such that

$$F^p(K) \bigcap F^q(K) = \emptyset, \quad |p - q| > \nu. \tag{2.2.7}$$

For the sake of definiteness we assume $F(x) > x$ and consider the interval $K = [x_0, F(x_0)]$ where $x_0 \in \mathbb{R}$ is an arbitrary point. It is evident that the set $V_p = \bigcup_{n \geq p} F^n(K)$ coincides with some semi-axis $[d_p, +\infty)$. Let us prove that $F(x)$ is strictly monotonic on V_p for sufficiently large p.

If the opposite statement is true, then for every $p \in \mathbb{N}$ there exist $n_p \geq p$, $m_p \geq p$, $a_p \in K$, $b_p \in K$, such that $F^{n_p}(a_p) \neq F^{m_p}(b_p)$, but $F^{n_p+1}(a_p) = F^{m_p+1}(b_p)$. If $|m_p - n_p| \to +\infty$ as $p \to \infty$, then since

$$F^{n_p+1}(K) \bigcap F^{m_p+1}(K) \neq \emptyset,$$

we arrive at a contradiction with (2.2.7). It means that we have to assume $|m_p - n_p| \leq t$ for all p. Let us check that this is impossible as well.

To this end we note that the set

$$K_1 = \bigcup_{0 \leq l \leq t} F^l(K)$$

is a closed bounded interval and there exists a number $\nu_1 = \nu(K_1)$ such that

$$F^l(K_1) \bigcap F^s(K_1) = \emptyset, \quad |l - s| \geq \nu_1.$$

To be definite we assume $m_p \geq n_p$ and set $c_p = F^{m_p - n_p}(b_p)$. It is evident that $a_p \in K_1$, $c_p \in K_1$, and

$$F^{n_p}(a_p) \neq F^{n_p}(c_p), \quad F^{n_p+1}(a_p) = F^{n_p+1}(c_p).$$

Since F has no periodic points, we have

$$F^l(a_p) \neq F^s(c_p), \quad l, s \leq n_p.$$

Of course, we can assume that n_p is increasing fast enough for the relation

$$n_p - 2n_{p-1} \to \infty$$

to be true. It follows now that the sets

$$W_1 = \{x : x = F^l(a_p), \ \nu_1 + n_{p-1} \leq l \leq n_p, \ p = 1, 2, \dots\}$$

and

$$W_2 = \{x : x = F^s(c_p), \ \nu_1 + n_{p-1} \leq s \leq n_p, \ p = 1, 2, \dots\}$$

do not intersect. Let us show that neither of them has finite limit points, implying that both sets are closed. Indeed, let

$$\lim_{j\to\infty} F^{l_j}(a_{p_j}) = z, \quad \lim_{j\to\infty} a_{p_j} = u.$$

Then choosing a neighborhood V of the interval with end-points z and u we get

$$F^{l_j}(V) \bigcap V \neq \emptyset$$

for all large j. The latter relation contradicts the absence of non-wandering compact sets. The same arguments show that W_2 is closed as well. Since $W_1 \bigcap W_2 = \emptyset$, there exists a continuous function $\gamma : \mathbb{R} \to [0,1]$ such that

$$\gamma\,|_{W_1} = 1, \quad \gamma\,|_{W_2} = 0.$$

If the cohomological equation with such γ has a continuous solutions φ, then

$$\varphi(F^{n_p+1}(a_p)) - \varphi(a_p) = \sum_{j=0}^{n_p} \gamma(F^j(a_p))$$

and

$$\varphi(F^{n_p+1}(c_p)) - \varphi(c_p) = \sum_{j=0}^{n_p} \gamma(F^j(c_p)).$$

Taking into account the values of γ on W_1 and W_2, and the identity $F^{n_p+1}(a_p) = F^{n_p+1}(c_p)$, we obtain

$$\varphi(c_p) - \varphi(a_p) = n_p - n_{p-1} - \nu_1 + 1 + \sum_{j=0}^{n_{p-1}-1+\nu_1} \left[\gamma(F^j(a_p)) - \gamma(F^j(c_p))\right].$$

Therefore

$$|\varphi(c_p) - \varphi(a_p)| \geq n_p - 2n_{p-1} - 2\nu_1 + 2 \to \infty.$$

The latter is not consistent with boundedness of φ on K_1 and the necessity of conditions $i)$ and $ii)$ is proved.

SUFFICIENCY. Under the same assumption $F(x) > x$ consider a semi-axis $[c, +\infty)$ such that

$$F : (c, +\infty) \to (F(c), +\infty)$$

is a homeomorphism and fix a number $d > F(c)$. Given a continuous function γ, consider a decomposition

$$\gamma(x) = \gamma_+(x) + \gamma_-(x)$$

such that

$$\gamma_+(x) = 0, \ x \geq d; \quad \gamma_-(x) = 0, \ x \leq F(c).$$

It is sufficient to construct continuous solutions $\varphi_\pm$ of equations

$$\varphi_\pm(F(x)) - \varphi_\pm(x) = \gamma_\pm(x). \tag{2.2.8}$$

Then the sum $\varphi = \varphi_+ + \varphi_-$ is a solution of (2.2.6). If F is a homeomorphism of the real line onto itself, we can set

$$\varphi_-(x) = \sum_{n=1}^{\infty} \gamma_-(F^{-n}(x)), \quad \varphi_+(x) = -\sum_{n=0}^{\infty} \gamma_+(F^n(x)). \tag{2.2.9}$$

For every fixed x both series have only a finite number of non-vanishing terms and this number is bounded in every finite interval containing x. Therefore the functions $\varphi_-(x)$ and $\varphi_+(x)$ are continuous and a straightforward calculation shows that they are solutions of (2.2.8).

We can use the same definition of $\varphi_+(x)$ for arbitrary mapping F, but since in the general case F may not be invertible, we need to change the definition of $\varphi_-(x)$ given by (2.2.9).

To this end we note that if $x \geq F^n(c)$, then there exists a unique $y \geq c$ such that $F^n(y) = x$, and we can define a system of continuous functions $\{ h_n(x) \}_{n=0}^{\infty}$ by setting $h_0(x) = \gamma_-(x)$, $x \in \mathbb{R}$, and

$$h_n(x) = \begin{cases} h_{n-1}(y), & x \geq F^n(c),\ x = F^n(y),\ y \geq c, \\ 0, & x \leq F^n(c), \end{cases}$$

for $n \geq 1$. A simple calculation shows that $h_n(F(x)) = h_{n-1}(x)$, $n \geq 1$. Instead of (2.2.9) we set

$$\varphi_-(x) = \sum_{n=1}^{\infty} h_n(x)$$

and obtain, in addition to $\varphi_+(x)$, a continuous solution $\varphi_-(x)$ of (2.2.8), completing the proof of Theorem 2.3. □

The following example shows that conditions of Theorem 2.3, being necessary for the solvability of the cohomological equation with arbitrary $\gamma(x)$, are not necessary for the solvability of the Abel equation and hence for the semi-conjugacy of F with the standard shift.

Example 2.3. For every integer $l \geq 1$ we consider two sequences $\{a_n^{(l)}\}_{n=1}^{\infty}$ and $\{b_n^{(l)}\}_{n=1}^{\infty}$ satisfying conditions

$$l < \cdots < a_{l+2}^{(l)} < b_{l+2}^{(l)} < a_{l+1}^{(l)} < b_{l+1}^{(l)} < a_l^{(l)} = b_l^{(l)} < \cdots < a_1^{(l)} = b_1^{(l)} = l + \frac{1}{2}$$

and

$$\lim_{n \to \infty} a_n^{(l)} = l.$$

For example, we can set

$$a_n^{(l)} = l + \frac{1}{2n}, \quad b_n^{(l)} = \begin{cases} l + \frac{1}{2n-1}, & 1 \le l \le n-1, \\ l + \frac{1}{2n}, & l \ge n. \end{cases}$$

Let $\tilde{F}$ be the piecewise linear mapping of $(1, +\infty)$ defined by the interpolation data

$$\tilde{F}(a_n^{(l)}) = a_n^{(l+1)}, \tilde{F}(b_n^{(l)}) = b_n^{(l+1)}, \ n = 1, 2, \ldots; \tilde{F}(l) = l+1, l = 1, 2, \ldots.$$

The continuous function

$$\psi(x) = \begin{cases} l, & l \le x \le l + 2^{-1}, \\ 2x - (l+1), & l + 2^{-1} \le x \le l+1, \end{cases}$$

satisfies the Abel equation $\psi(\tilde{F}(x)) - \psi(x) = 1$, $x \in (1, +\infty)$. The homeomorphism $\Phi(x) = \ln(x-1)$, $x \in (1, +\infty)$, transplants $\tilde{F}$ on the entire real line and yields the mapping

$$F(x) = \ln \tilde{F}(e^x + 1) - 1, \quad x \in \mathbb{R}.$$

Since ψ satisfies the Abel equation for $\tilde{F}$, the function $\varphi(x) = \psi(e^x + 1)$ semi-conjugates F with the standard shift.

On the other hand,

$$a_n^{(n-1)} = n - 1 + (2n)^{-1} < b_n^{(n-1)} = n - 1 + (2n-1)^{-1},$$

but

$$\tilde{F}(a_n^{(n-1)}) = \tilde{F}(b_n^{(n-1)}) = n + (2n)^{-1}.$$

Therefore F is not strictly monotonic on every semi-axis $(c, +\infty)$ and according to Theorem 2.3 there exist continuous functions γ such that equation (2.2.6) has no continuous solutions on $\mathbb{R}$.

2.2.3 Smooth and analytic solutions of a cohomological equation

We discuss here the existence of smooth solutions of equation (2.2.6) with a smooth mapping F.

Theorem 2.4. *Let $F : \mathbb{R} \to \mathbb{R}$ be a C^k- mapping without fixed and critical points. Then the cohomological equation (2.2.6) has a C^k-solution for every $\gamma(x)$ from the same smoothness class.*

Proof. We repeat, essentially, the proof of Theorems 2.1 and 2.2. Let, as before, x_0 be either an arbitrary point if $\lim_{x \to -\infty} F(x) = -\infty$ or $x_0 = \lim_{x \to -\infty} F(x)$ if the latter limit is finite, and let $\varphi_0(x)$ be a local C^k-solution in a neighborhood of $[x_0, F(x_0)]$. For a finite k and $k = \infty$ such a solution was described in the

Preface. Its construction for analytic functions is postponed until Section 3.1.1 (see Corollary 3.1, Chapter 3). Then the formula

$$\varphi_1(x) = \varphi_0(F^{-n}(x)) + \sum_{k=1}^{n} \gamma(F^{-k}(x)), \quad F^n(x_0) \le x \le F^{n+1}(x_0), \ n \in \mathbb{N},$$

defines a C^k-function on the semi-axis $[x_0, +\infty)$. If F is surjective, we can extend the previous definition to the entire axis by (2.2.6). Since the mapping $F(x)$ has no critical points, this extension defines a C^k-solution of (2.2.6).

In the case $\lim_{x\to-\infty} F(x) > -\infty$ we set

$$\varphi(x) = \begin{cases} \varphi_1(x), & x \ge x_0, \\ \varphi_0(F(x)) - \gamma(x), & x \le x_0, \end{cases}$$

and again obtain a solution from the same class as $\gamma(x)$. □

The following example shows that the cohomological equation may not have smooth solutions in the presence of critical points.

Example 2.4. For every α, $0 < \alpha < 1$, the mapping

$$F_\alpha(x) = x + 1 - \alpha \sin \frac{x}{\alpha}, \quad x \in \mathbb{R},$$

is a real-analytic homeomorphism without fixed points. According to Theorem 2.3 the corresponding cohomological equation has a continuous solution for every continuous function $\gamma(x)$. The set of critical points of F_α is

$$C(F_\alpha) = \{ x : F'_\alpha(x) = 0 \} = \{ 2\pi s\alpha \}_{s=-\infty}^{\infty}.$$

We will show that if α is chosen such that

$$F_\alpha^n(C(F_\alpha)) \bigcap C(F_\alpha) = \emptyset, \quad \forall n \ne 0, \tag{2.2.10}$$

then there exists an analytic function $\gamma(x)$ such that the cohomological equation (2.2.6) has no C^1-solutions.

Since

$$F_\alpha^n(2\pi s\alpha) = F_\alpha^n(0) + 2\pi s\alpha, \quad n, s = 0, \pm 1, \dots,$$

condition (2.2.10) means that $F_\alpha^n(0) \ne 2s\pi\alpha$ for all $n, s \in \mathbb{Z}$. In other words, α satisfies (2.2.10) if and only if it does not belong to the set $\bigcup Z_{n,s}$ where $Z_{n,s}$ is the zero set of the analytic function $\theta_{n,s}(\alpha) = F_\alpha^n(0) - 2s\pi\alpha$. It follows from this that the set of such α is uncountable.

To construct an appropriate function γ we note that if φ is a smooth solution of (2.2.6), then iterating and differentiating the cohomological equation, we obtain

$$\varphi'(x) = a_{n+1}(x)\varphi'(F_\alpha^{n+1}(x)) - \sum_{k=0}^{n} a_k(x)\gamma'(F_\alpha^k(x))$$

where

$$a_k(x) = (F_\alpha^k(x))' = \prod_{j=0}^{k-1} F_\alpha'(F^j(x)).$$

If we set here $x = F_\alpha^{-n}(x_s)$ where $x_s = 2\pi s\alpha$ is a critical point of F_α, then we obtain $a_{n+1}(x) = 0$ and hence

$$\varphi'(F_\alpha^{-n}(x_s)) = -\sum_{k=0}^{n} a_k(F_\alpha^{-n}(x_s))\gamma'(F_\alpha^{k-n}(x_s)), \tag{2.2.11}$$

where $a_n(F_\alpha^{-n}x_s) \neq 0$ because of (2.2.10). Set

$$s_n = E\left[-\frac{1}{2\pi\alpha}F_\alpha^{-n}(0)\right]$$

where $E[x]$ is the entire part of $x \in \mathbb{R}$. Obviously, $s_n \to +\infty$ and

$$|F_\alpha^{-n}(x_{s_n})| = |x_{s_n} + F_\alpha^{-n}(0)| \leq 2\pi\alpha.$$

Now let γ be a function satisfying inequalities

$$|\gamma'(x_{s_n})| \geq \frac{1}{|a_n(F_\alpha^{-n}(x_{s_n}))|}\left(n + \sum_{k=0}^{n-1}|a_k(F_\alpha^{-n}(x_{s_n}))\gamma'(F_\alpha^{k-n}(x_{s_n}))|\right).$$

According to (2.2.11) every smooth solution $\varphi(x)$ of (2.2.6) satisfies the inequality

$$|\varphi'(F_\alpha^{-n}(x_{s_n}))| \geq n, \quad \forall n \in \mathbb{N}.$$

It means that $\varphi'(x)$ is not bounded in $[-2\pi\alpha, 2\pi\alpha]$ which is impossible. Hence (2.2.6) has no smooth solutions for such $\gamma(x)$.

Let us note that condition (2.2.10) does not hold, for example, if $\alpha = \pi^{-1}$. It may be shown that in the non-analytic situation the corresponding cohomological equation has a C^k-solution for every C^k-function γ.

2.3 Local structure of mappings at an isolated fixed point

The first step in a study of continuous mappings with a non-empty set of non-wandering points on M is to investigate their local behavior. Due to the local nature of the problem it is sufficient to consider such behavior in a neighborhood of a point on the real line $\mathbb{R}$.

Definition 2.2. Two mappings F and G defined in a neighborhood V of a common invariant set S are said to be *locally* C^k*-conjugate* at S if there exist a neighborhood $U \subset V$ of S and C^k-diffeomorphism

$$\Phi : U \to \Phi(U), \quad \Phi(U) \subset V, \quad \Phi(S) = S,$$

such that

$$\Phi(F(x)) = G(\Phi(x)), \quad x \in U \bigcap F^{-1}(U).$$

A fixed point of F is the simplest non-trivial invariant set. In a neighborhood U of isolated fixed point x_0 there are the following patterns of dynamical behavior:

a) $F^n(x) \to x_0$, $n \to +\infty$, $x \in U$, and x_0 is called an *attractor.*

b) $F^{-n}(x) \to x_0$, $n \to +\infty$, $x \in U$, and x_0 is called a *repeller.* In other words, a repeller of F is an attractor for F^{-1}.

c) $F^n(x) \to x_0$, $n \to +\infty$, $x \leq x_0$, $x \in U$, and $F^{-n}(x) \to x_0$, $n \to +\infty$, $x \geq x_0$, $x \in U$, and x_0 is called an *overflow from left to right.* If x_0 is an overflow from left to right for F^{-1}, it is called an *overflow from right to left* for F; both types of overflows are conjugate by $\Phi(x) = -x$.

An example of mappings of types a) and b) is given by a linear function

$$F(x) = \lambda x, \quad x_0 = 0.$$

Here $x_0 = 0$ is an attractor if $|\lambda| < 1$ and a repeller if $|\lambda| > 1$. The function

$$F(x) = \frac{x}{1 + x}, \quad x_0 = 0,$$

supplies an example of a mapping with an overflow from right to left: the point $x_0 = 0$ attracts all points $x \geq 0$ and repels all points $x \in (-1, 0]$.

It easily follows from Theorem 2.1 that two orientation-preserving homeomorphisms are topologically conjugate in a neighborhood of a common isolated fixed point if and only if they belong to the same type of local behavior. Thus, the dynamical type is the unique invariant of the local topological classification in a neighborhood of an isolated fixed point.

The situation is different for the smooth classification. Namely, if two smooth mappings F and G are smoothly conjugate at a common fixed point x_0, then necessarily $F'(x_0) = G'(x_0)$, and hence the number $F'(x_0)$ is an invariant of smooth classification. The following example shows that, generally speaking, the equality of the derivatives is not sufficient for F and G to be locally smoothly conjugate.

Example 2.5. Let, for a fixed $\lambda \in (0, 1)$,

$$F(x) = \begin{cases} \lambda x(1 + (\ln |x|)^{-1}), & 0 < |x| < 1, \\ 0, & x = 0. \end{cases}$$

Then $F \in C^1$, $F'(0) = \lambda$, but, as we will show now, $F(x)$ is not smoothly conjugate with $G(x) = \lambda x$. Indeed, if there exists a conjugating diffeomorphism $\Phi(x)$, then in a small neighborhood of $x_0 = 0$ the estimate $c^{-1}|x| \le |\Phi(x)| \le c|x|$ holds with $c > 0$. Iterating the equation $\Phi(F(x)) = \lambda\Phi(x)$ we obtain $\Phi(F^n(x)) = \lambda^n\Phi(x)$ and

$$1 \le c^2\lambda^{-n}\left|\frac{F^n(x)}{x}\right| = c^2\prod_{k=1}^{n}\lambda^{-1}\left|\frac{F(F^{k-1}(x))}{F^{k-1}(x)}\right|$$

$$= c^2\prod_{k=1}^{n}\left(1 + \frac{1}{\ln|F^{k-1}(x)|}\right).$$

Since

$$\sum_{k=1}^{\infty}\frac{1}{\ln|F^{k-1}(x)|} = -\infty, \quad 0 < |x| < 1,$$

the infinite product tends to 0, and we arrive at a contradiction.

According to an accepted definition a fixed point x_0 of a mapping $F \in C^1$ is called *hyperbolic* if $|F'(x_0)|$ is neither 0 nor 1. In Example 2.5 we constructed C^1 (but not C^2)-local diffeomorphism F with a hyperbolic fixed point x_0, which is not locally smoothly conjugate with $G(x) = x_0 + F'(x_0)(x - x_0)$. The following statement shows that in classes of higher smoothness the number $F'(x_0)$ is the unique invariant with respect to conjugacy transformations at a hyperbolic fixed point.

Theorem 2.5. *Let F be a local C^k-diffeomorphism, $k \ge 2$, hyperbolic at a fixed point x_0. Then F is C^k-locally conjugate at x_0 with the linear mapping $G(x) = x_0 + F'(x_0)(x - x_0)$.*

Proof. Without loss of generality we assume $x_0 = 0$. We restrict ourselves to the case $|\lambda| < 1$; otherwise instead of F we consider the inverse diffeomorphism F^{-1}. The problem now is to prove the local solvability of a special Schröder equation

$$\Phi(F(x)) = \lambda\Phi(x), \quad \lambda = F'(0),$$

where Φ must be a local diffeomorphism. If we set $\Phi(x) = x + \varphi(x)$, $\varphi(0) = \varphi'(0) = 0$, then we obtain the equation

$$\varphi(F(x)) - \lambda\varphi(x) = \gamma(x), \quad \varphi(0) = \varphi'(0) = 0, \tag{2.3.1}$$

where

$$\gamma(x) = \lambda x - F(x).$$

For every $\varepsilon > 0$ there exist sufficiently small $\delta(\varepsilon) > 0$ and some $C > 0$ such that

$$|F(x)| \le (|\lambda| + \varepsilon)|x|, \quad |\gamma(x)| = |\lambda x - F(x)| \le Cx^2, \quad |x| \le \delta(\varepsilon).$$

If $|\lambda| + \varepsilon < 1$, then $|F^n(x)| \leq (|\lambda| + \varepsilon)^n |x|$ for all $n \geq 1$, and iterating (2.3.1) we find that the unique smooth solution of (2.3.1) is given by the series

$$\varphi(x) = -\sum_{n=0}^{\infty} \lambda^{-n-1}\gamma(F^n(x)).$$

Let us show that in fact $\varphi(x)$ belongs to the same class C^k as $\gamma(x)$.

To this end we fix $\varepsilon > 0$, $\delta > 0$ such that

$$q = |\lambda|^{-1}(|\lambda| + \varepsilon)^2 < 1, \quad |F(x)| \leq (|\lambda| + \varepsilon)|x|, \quad |F'(x)| \leq |\lambda| + \varepsilon,$$

for $|x| \leq \delta$. For $2 \leq k < \infty$ consider the subspace $\mathcal{L} \subset C^k[-\delta, \delta]$ of all C^k-functions ψ such that

$$\|\psi\|_0 = \max_{|x|\leq\delta} \frac{|\psi(x)|}{|x|^2} < \infty, \quad \|\psi\|_1 = \max_{|x|\leq\delta} \frac{|\psi'(x)|}{|x|} < \infty.$$

Obviously, $\gamma \in \mathcal{L}$. It is sufficient to show that the operator

$$T\psi(x) = \lambda^{-1}\psi(F(x)), \quad \psi \in \mathcal{L},$$

is contractive in the space $\mathcal{L}$ endowed with an appropriate norm. To define such a norm let us set

$$\|\psi\|_j = \max_{|x|\leq\delta} |\psi^{(j)}(x)|, \quad 2 \leq j \leq k.$$

Then

$$\|T\psi\|_0 \leq q\|\psi\|_0, \quad \|T\psi\|_1 \leq q\|\psi\|_1,$$

and

$$\|T\psi\|_j \leq q\|\psi\|_j + M_j \sum_{i=1}^{j-1} \|\psi\|_i, \quad j = 2, 3, \ldots, k,$$

where numbers M_j depends on F only. Choose $\theta > 0$ such that $q + \theta < 1$ and set

$$b = 2 + \frac{1}{\theta}\max(M_1, M_2, \ldots, M_k).$$

For

$$\|\psi\| = \sum_{j=0}^{k} b^j \|\psi\|_{k-j}$$

we obtain

$$\|T\psi\| = \sum_{j=0}^{k} b^j \|T\psi\|_{k-j} \leq q\|\psi\| + \sum_{j=0}^{k-2} M_{k-j} b^j \sum_{i=1}^{k-j-1} \|\psi\|_i$$

$$\leq q\|\psi\| + \theta \sum_{j=1}^{k-1} b^{j-1} \sum_{i=1}^{k-j} \|\psi\|_i$$

$$\leq q\|\psi\| + \theta \sum_{j=1}^{k-1} \|\psi\|_i \sum_{j=1}^{k-i} b^{j-1} \leq (q + \theta)\|\psi\|.$$

Therefore operator T is a contraction in $\mathcal{L}$ and the function $\varphi(x) = \sum_{n=0}^{\infty} T^n\gamma(x)$ belongs to $\mathcal{L}$.

If $F \in C^\infty$, then, as we have already shown, $\varphi \in C^k$ for every $k \geq 2$, and hence $\varphi \in C^\infty$ as well.

In the analytic case we consider the Banach space of all functions ψ analytic in a disc $D = \{ z : |z| < \delta \} \subset \mathbb{C}$, continuous up to its boundary and having finite norm

$$\|\psi\|_0 = \max_{z\in D}\left|\frac{\psi(z)}{z^2}\right|.$$

As before, we have $\|T\psi\|_0 \leq q\|\psi\|_0$ which implies that the unique solution of (2.3.1) is analytic at $x = 0$.

This completes the proof of Theorem 2.5 for all smoothness classes. □

In what follows we will use similar constructions to prove the smooth local solvability of more general equations.

Non-hyperbolic cases $\lambda = 0$ and especially $\lambda = \pm 1$ are more complicated. In any case, if $F(x)$ is conjugate with either $G(x) \equiv 0$ or $G(x) \equiv x$, then either $F(x) \equiv 0$ or $F(x) \equiv x$. It means that in the non-hyperbolic situation a linear function $G(x) = cx$ cannot be a "normal" form for a smooth mapping. Our nearest aim is to find an appropriate normal form at a non-hyperbolic fixed point.

2.3.1 Formal classification

Let F and G be two local C^∞-mapping, $F(0) = G(0) = 0$, which are locally conjugate by a C^∞-diffeomorphism Φ, $\Phi(0) = 0$. Then the formal Taylor series $\hat{\Phi}$ of Φ satisfies the *formal* Schröder equation

$$\hat{\Phi}(\hat{F}(x)) = \hat{G}(\hat{\Phi}(x))$$

where $\hat{F}$ and $\hat{G}$ are the formal Taylor series at $x = 0$ of F and G, respectively.

We denote by $\mathbb{R}[[x]]$ the set of all formal series in x with real coefficients and zero coefficient at x^0. The set $\mathbb{R}[[x]]$ is an algebra over the field $\mathbb{R}$ and, besides, a semi-group with respect to a substitution. The unit series $I(x) \equiv x$ is the identity of the semi-group, and the group of invertible elements is formed by all series with non-zero coefficient at x. Two formal series $\hat{F}$ and $\hat{G}$ from $\mathbb{R}[[x]]$ are called *conjugate* if they are conjugate with respect to the group of invertible elements, i.e., if $\hat{G} = P^{-1}\hat{F}P$ with some invertible series P.

Theorem 2.6. *Every formal series from $\mathbb{R}[[x]]$ is conjugate with one of the following pairwise non-conjugate normal forms:*

I. $G(x) = \pm x^r, \quad r \geq 2;$

II. $G(x) = \lambda x;$

III. $G(x) = x \pm x^r + \alpha x^{2r-1}, \quad r \geq 2;$

IV. $G(x) = -x \pm x^{2r+1} + \alpha x^{4r+1}, \quad r \geq 1.$

Proof. The proof is reduced to successive "killing" of unnecessary terms using transformations

$$\Phi_p(x) = x + \mu x^p, \quad p = 2, 3, \dots, \tag{2.3.2}$$

and the resulting transformation is an infinite product

$$\hat{\Phi} = \Phi_2 \Phi_3 \Phi_4 \dots$$

which converges in $\mathbb{R}[[x]]$.

Let us begin with the degenerate case $\hat{F}(x) = c_2 x^2 + \cdots$. If all coefficients of $\hat{F}$ are zero, then $\hat{F}$ coincides with $G(x) = 0$ which is the normal form II with $\lambda = 0$. Otherwise, let

$$\hat{F}(x) = cx^r + \cdots, \quad c \neq 0, \ r \geq 2.$$

A linear transformation $\Phi(x) = \mu x$ reduces the initial term cx^r to the form $\pm x^r$. If

$$\hat{F}(x) = \pm x^r + cx^p + \cdots, \quad p \geq r+1,$$

then we set $\Phi(x) = x + \mu x^{p-r+1}$ and obtain

$$\Phi^{-1} \hat{F} \Phi = \pm x^r \pm r\mu x^p + cx^p + \cdots.$$

With $r\mu = \mp c$ the latter series does not contain the term with x^p. Repeating the same procedure we arrive at the normal form I.

Let now $\hat{F}(x) = \lambda x + f_p x^p + \cdots$, $\lambda \neq 0$, $p \geq 2$. If $\hat{\Phi}(x) = \Phi_p(x)$ is defined by (2.3.2), then $\hat{\Phi}^{-1}(x) = x - \mu x^p + \cdots$ and

$$\hat{\Phi}^{-1} \hat{F} \hat{\Phi}(x) = \hat{\Phi}^{-1}(\lambda x + \lambda \mu x^p + f_p x^p + \cdots) =$$

$$= \lambda x + (\lambda \mu + f_p - \mu \lambda^p) x^p + \cdots.$$

If $\lambda^p \neq \lambda$ for all $p \geq 2$, then we set

$$\mu = \frac{f_p}{\lambda^p - \lambda}$$

and successively applying (2.3.2) obtain a linear normal form II.

The equation $\lambda^p = \lambda$ is the simplest version of *resonance relations* well-known in local dynamics. Such relations are obstacles to reducing a mapping to a linear form.

The resonance relation $\lambda^p = \lambda$ is possible if and only if $\lambda = \pm 1$. If $\lambda = -1$, then successively applying transformations (2.3.2) we annihilate all terms with x^{2p} and reduce the series $\hat{F}$ to a *resonance normal form*

$$\hat{F}(x) = -x + \sum_{p=1}^{\infty} c_p x^{2p+1}$$

corresponding to resonance relations $\lambda^{2p+1} = \lambda$.

If $\hat{F}(x) \not\equiv -x$ (otherwise $\hat{F}$ is of the form II with $\lambda = -1$), we can assume

$$\hat{F}(x) = -x \pm x^{2r+1} + \cdots .$$

Consecutive transformations

$$\Phi(x) = x + \mu x^{2p-2r+1}, \quad p > r,$$

annihilate all terms $c_p x^{2p+1}$ with $p \neq 2r$ and reduce $\hat{F}$ to the form IV.

For $\lambda = 1$ all exponents p are resonant. If $\hat{F}(x) = x$, then the series and $\hat{F}$ has the normal form II with $\lambda = 1$. Otherwise

$$\hat{F}(x) = x + cx^r + \cdots$$

with some $c \neq 0$, $2 \leq r < \infty$. A linear transformation $\Phi(x) = \mu x$ reduces $\hat{F}$ to a form

$$\hat{F}(x) = x \pm x^r + bx^p + \cdots .$$

If $\hat{F}(x) \equiv x \pm x^r$, then we have the normal form III with $\alpha = 0$. To get rid of bx^p with $b \neq 0$ we set

$$\Phi(x) = x + \mu x^{p-r+1}.$$

Then

$$\Phi^{-1}\hat{F}\Phi(x) = x \pm x^r + (\pm(2r - p - 1)\mu + b)x^p + \cdots$$

which means that we can annihilate terms of all orders $p \neq 2r - 1$ and arrive at the normal form III. □

Let us note that the parameter α entering the normal forms III and IV is an invariant: if $\hat{F}$ and $\hat{G}$ are conjugate in $\mathbb{R}[[x]]$, then the corresponding values of this parameter coincide.

As we will see below, it is more convenient for some applications to use "model" mappings different from I–IV. For instance, Theorem 2.6 implies

Corollary 2.3. *Every formal diffeomorphism $\hat{F}(x)$ is conjugate either with $G'(x) = \lambda x$ or with the formal Taylor series of a mapping*

$$G(x) = \frac{\pm x}{(1 \pm rx^r + \beta x^{2r} + x^{2r}g(x))^{1/r}}, \quad r \geq 1, \ g(0) = 0,$$

with an arbitrary function $g \in C^\infty$, and the unique choice of r, β and signs in the numerator and denominator.

Proof. To check the statement we note that

$$G(x) = \pm x \left(1 \mp x^r + \left(\frac{r+1}{2} - \frac{\beta}{r}\right) x^{2r} + \cdots \right).$$

Now it is sufficient to choose appropriate values of parameters r and β. □

One of the advantages of such mappings is the relative simplicity of their iterations. For example, if

$$G_{\pm}(x) = \frac{\pm x}{(1 + \mu x^r)^{1/r}},$$

then formally

$$G_+^n(x) = \frac{x}{(1 + n\mu x^r)^{1/r}}, \quad n \in \mathbb{Z},$$

and

$$G_-^n(x) = (-1)^n \frac{x}{(1 + n\mu x^r)^{1/r}}, \quad n \in \mathbb{Z},\ r = 2k.$$

For $r = 2k+1,\ k \in \mathbb{N},$ we obtain $G_-^2(x) \equiv x.$

If $\mu > 0$ and $r = 2k,$ then the iterations $G_+^n(x)$ are defined for all $n > 0$ and all $x \in \mathbb{R}$. It is evident that

$$n^{\frac{1}{r}} G_+^n(x) \to \mu^{-\frac{1}{r}}, \quad n \to +\infty,\ x > 0.$$

Let now

$$F(x) = G_+(x) + o(x^{r+1}), \quad x \to 0.$$

Since $G_+(x) = x(1 - \mu r^{-1} x^r) + o(x^{r+1}),$ for all sufficiently small positive δ and ε the inequalities

$$\frac{x}{(1 + (\mu + \delta)x^r)^{1/r}} \le F(x) \le \frac{x}{(1 + (\mu - \delta)x^r)^{1/r}}, \quad 0 < x < \varepsilon,$$

hold. The function $F(x)$ is increasing in $\{x : |x| \le \varepsilon\}$ and these inequalities may be iterated yielding

$$\frac{x}{(1 + n(\mu + \delta)x^r)^{1/r}} \le F^n(x) \le \frac{x}{(1 + n(\mu - \delta)x^r)^{1/r}}, \quad 0 < x < \varepsilon.$$

Therefore

$$n^{\frac{1}{r}} F^n(x) \to \mu^{-\frac{1}{r}}, \quad n \to +\infty,\ x > 0. \tag{2.3.3}$$

2.3.2 Smooth classification

Given a smooth mapping F, it is natural to expect that it may be reduced to a normal form. The following proposition shows that such an expectation is justified at least for "almost all" mappings from C^∞.

Theorem 2.7. *Let F and G be two C^∞-mappings with a common fixed point $x = 0$, and let neither $\hat{F}(x) \equiv 0$ nor $\hat{F}^2(x) \equiv x$. If the formal series $\hat{F}$ and $\hat{G}$ are conjugate at $x = 0$, then the mappings F and G are locally C^∞-conjugate as well.*

The statement of Theorem 2.7 implies that the equivalence class of a mapping F satisfying its conditions is completely determined by the formal normal form. Namely,

Corollary 2.4. *Every local C^∞-mapping F with a fixed point $x = 0$ for which neither $\hat{F}(x) \equiv 0$ nor $\hat{F}^2(x) \equiv x$ is locally C^∞-conjugate with one of normal forms* I–IV.

According to Theorem 2.5, a similar situation occurs for C^k-smooth diffeomorphisms with $k \geq 2$ at hyperbolic fixed points. But in a general case, as we have already seen in Example 2.5 above, the classification based on the reduction to a formal normal form is not possible. It means that in addition to a formal normal form there exist some other invariants of the classification, and here we propose some approach to their description. We postpone the proof of Theorem 2.7 until the end of the present section where it will be obtained as an immediate result of this approach.

Lemma 2.1. *Let F be a C^k-diffeomorphism in a neighborhood of the isolated fixed point $x = 0$ such that the semi-axes $(-\infty, 0]$ and $[0, +\infty)$ are locally invariant with respect to F. Then there exists C^k-diffeomorphism Ψ_+ of a right semi-neighborhood $(0, \varepsilon)$ on a neighborhood of either $+\infty$ or $-\infty$, which conjugates F with the standard shift. The choice of $\pm\infty$ depends on either the point $x = 0$ attracts or repels on $(0, \varepsilon)$.*

Proof. To be definite we assume $x = 0$ to be a point attracting from the right and consider the diffeomorphism

$$F_\infty(t) = \frac{1}{F(t^{-1})}$$

defined at a neighborhood of ∞. Then $F_\infty(t) > t$ for large t, and since $F(t)$ is fixed-point free, we can extend $F_\infty(t)$ to a fixed-point free diffeomorphism of the entire real axis from the same class C^k for finite k and $k = \infty$. According to Theorem 2.2 there exists an orientation-preserving C^k-diffeomorphism Φ conjugating F_∞ with the standard shift, i.e., $\Phi(F_\infty(t)) = \Phi(t) + 1$. To complete the proof we set

$$\Psi_+(x) = \Phi(x^{-1}), \quad x \in (0, \varepsilon).$$

If $F_\infty(t)$ may be continued as an analytical mapping without fixed and critical points on the entire real axis, then according to the same Theorem 2.2 it is analytically conjugate with the standard shift, and Ψ_+ may be defined by the latter formula, Φ being a conjugating diffeomorphism. Generally speaking, $F_\infty(t)$ is not analytically continuable to the entire real axis, but nevertheless an analytic diffeomorphism Φ exists and will be constructed later in this section (see Lemma 2.3 below). □

Let us note that every diffeomorphism F from Lemma 2.1 preserves the orientation. Similar to the statement of Lemma 2.1, there exists a C^k-diffeomorphism Ψ_- defined on $(-\varepsilon, 0)$ and conjugating F with the standard shift either at $+\infty$ or $-\infty$. Therefore to every local orientation-preserving diffeomorphism F with an isolated fixed point x_0 there corresponds a pair of diffeomorphisms $\Psi_\pm$ of the same smoothness as F such that the representation

$$F(x) = \begin{cases} \Psi_+^{-1}(\Psi_+(x) + 1), & x > x_0, \\ 0, & x = x_0, \\ \Psi_-^{-1}(\Psi_-(x) + 1), & x < x_0, \end{cases} \tag{2.3.4}$$

holds. We will call every such pair $\Psi_\pm$ the *straightening diffeomorphisms for F at x_0*.

Of course, diffeomorphisms $\Psi_\pm$ are not determined uniquely: if $H_\pm : \mathbb{R} \to \mathbb{R}$ is any pair of C^k-diffeomorphisms commuting with the standard shift, then the pair $\tilde{\Psi}_\pm = H_\pm \Psi_\pm$ is straightening for the same mapping F. On the other side, if two pairs $\tilde{\Psi}_\pm$ and $\Psi_\pm$ correspond to the same C^k-diffeomorphism F, then the diffeomorphisms $H_\pm = \tilde{\Psi}_\pm \Psi_\pm^{-1}$ commute with the standard shift.

All mappings commuting with the standard shift have the form

$$H(t) = t + \tau(t),$$

where $\tau(t)$ is a 1-periodic function. A mapping $H(t)$ is a C^k-diffeomorphism if and only if $\tau \in C^k$ and

$$1 + \tau'(t) > 0.$$

Example 2.6. The straightening diffeomorphisms for a linear mapping $F(x) = \lambda x$, $\lambda > 0$, $\lambda \neq 1$, have the form

$$\Psi_+(x) = \frac{1}{\ln \lambda} \ln x + \tau_+ \left(\frac{\ln |x|}{\ln \lambda} \right), \qquad x > 0,$$

$$\Psi_-(x) = \frac{1}{\ln \lambda} \ln |x| + \tau_- \left(\frac{\ln |x|}{\ln \lambda} \right), \qquad x < 0,$$

where $\tau_\pm$ are 1-periodic functions.

Example 2.7. The straightening diffeomorphisms for the mapping

$$G(x) = \frac{x}{(1 + r x^r)^{1/r}}, \qquad r = 1, 2, \ldots$$

have the form

$$\Psi_\pm(x) = \frac{1}{r} x^{-r} + \tau_\pm \left(\frac{1}{r} x^{-r} \right)$$

where $\tau_\pm$ are 1-periodic functions.

Generally speaking, not all pairs of diffeomorphisms $\Psi_\pm$ of a neighborhood of $x = 0$ onto a neighborhood of a point at infinity may be pairs of straightening diffeomorphisms for a local diffeomorphism F. Indeed, if Ψ_+ and Ψ_- are straightening diffeomorphisms of a C^1-diffeomorphism $F(x) = \lambda x + o(x)$, $\lambda \in (0,1)$, then

$$\Psi_\pm(F(x)) = \Psi_\pm(x) + 1,$$

and the relations

$$\lim_{t\to+\infty} \frac{(\Psi_+^{-1})'(t+1)}{(\Psi_+^{-1})'(t)} = \lim_{t\to+\infty} \frac{(\Psi_-^{-1})'(t+1)}{(\Psi_-^{-1})'(t)} = \lambda \tag{2.3.5}$$

necessarily hold where $\Psi_\pm^{-1}$ are the inverse diffeomorphisms mapping a neighborhood of a point at $+\infty$ onto the corresponding semi-neighborhood of $x = 0$. On the other hand, if some C^1-diffeomorphisms Ψ_+ and Ψ_- of $(0, +\varepsilon)$ and $(-\varepsilon, 0)$, respectively, onto a neighborhood of ∞, satisfy (2.3.5) with $\lambda \in (0,1)$, then the representation (2.3.4) defines a local C^1-diffeomorphism $F(x) = \lambda x + o(x)$ with the straightening diffeomorphisms Ψ_+ and Ψ_-.

Let us now derive a representation similar to (2.3.4) for an orientation-inverting diffeomorphism F.

Denote by $\Phi_\pm$ a pair of straightening diffeomorphisms for the orientation-preserving diffeomorphism F^2 and set

$$F_1(x) = \begin{cases} \Phi_-^{-1}(\Phi_+(x) + 2^{-1}), & x > 0, \\ \Phi_+^{-1}(\Phi_-(x) + 2^{-1}), & x < 0. \end{cases}$$

Then F_1 is a diffeomorphism of a punctured neighborhood of the origin, $F_1^2 = F^2$, but maybe F_1 has no smooth continuation to the point $x = 0$. Set now $\Psi_\pm = \Phi_\pm W^{-1}$ where

$$W(x) = \begin{cases} FF_1^{-1}(x), & x > 0, \\ x, & x < 0. \end{cases}$$

Then the diffeomorphisms $\Psi_\pm$ are straightening for F^2 as well and

$$F(x) = \begin{cases} \Psi_-^{-1}(\Psi_+(x) + 2^{-1}), & x > 0, \\ 0, & x = 0, \\ \Psi_+^{-1}(\Psi_-(x) + 2^{-1}), & x < 0, \end{cases} \tag{2.3.6}$$

Let us state a general criterion of conjugacy in terms of diffeomorphisms $\Psi_\pm$.

Theorem 2.8. *Let F and $\tilde{F}$ be two orientation-preserving diffeomorphisms of the same dynamical behavior at the common fixed point $x = 0$, and let $\Psi_\pm$ and $\tilde{\Psi}_\pm$ be corresponding straightening diffeomorphisms. Then for F and $\tilde{F}$ to be C^p-locally conjugate it is necessary and sufficient that there exist two C^p-diffeomorphisms $H_\pm$ commuting with the standard shift and a local C^p-diffeomorphism Φ, $\Phi(0) = 0$, such that*

$$\tilde{\Psi}_\pm(\Phi(x)) = H_\pm(\Psi_\pm(x)). \tag{2.3.7}$$

Proof. Let

$$\tilde{F}(x) = \Phi F \Phi^{-1}(x)$$

where Φ is a local C^p-diffeomorphism. Then the C^p-mappings

$$H_\pm(t) = \tilde{\Psi}_\pm \Phi \Psi_\pm^{-1}(t)$$

are well defined in a neighborhood of a point at infinity and commute with the standard shift. The latter relation is equivalent to (2.3.7).

On the other hand, if (2.3.7) is fulfilled, then the diffeomorphism Φ conjugates F and $\tilde{F}$. □

Conditions of Theorem 2.8 may be stated in the following form:

There exist diffeomorphisms $H_\pm$ commuting with the standard shift such that the mapping

$$\Phi(x) = \begin{cases} \tilde{\Psi}_+^{-1} H_+ \Psi_+(x), & x > 0, \\ 0, & x = 0, \\ \tilde{\Psi}_-^{-1} H_- \Psi_-(x), & x < 0, \end{cases} \qquad (2.3.8)$$

is a local C^p-diffeomorphism.

In turn this statement may be formulated in terms of a joint behavior at infinity of the mappings $\Psi_\pm$ and $\tilde{\Psi}_\pm$. For example, similar to (2.3.5), the C^1-conjugacy of diffeomorphisms F and $\tilde{F}$ with the attracting point $x = 0$ is equivalent to the existence of C^1-mappings $H_\pm$ commuting with the standard shift and such that there exist finite and equal limits

$$\lim_{t\to+\infty} \frac{(\tilde{\Psi}_-^{-1}(H_-(t))'H_-'(t)}{(\Psi_-^{-1}(t))'} = \lim_{t\to+\infty} \frac{(\tilde{\Psi}_+^{-1}(H_+(t))'H_+'(t)}{(\Psi_+^{-1}(t))'} \neq 0.$$

To obtain a conjugacy criterion for diffeomorphisms F and $\tilde{F}$ which do not preserve an orientation, we can use representation (2.3.6). Namely, let $\Psi_\pm$ and $\tilde{\Psi}_\pm$ be two pairs of straightening diffeomorphisms for F^2 and $\tilde{F}^2$, respectively, such that (2.3.6) holds. It follows from Theorem 2.8 that

Two orientation-inverting C^k-diffeomorphisms F and $\tilde{F}$ are locally C^p-conjugate if and only if there exist diffeomorphisms $H_\pm$ commuting with the standard shift and satisfying $H_+(t + 2^{-1}) = H_-(t) + 2^{-1}$ such that Φ defined by (2.3.8) is a local C^p-diffeomorphism.

Using the above scheme let us study in more detail the C^1-conjugacy at a hyperbolic fixed point.

Let $F(x) = \lambda x + o(x)$ be a C^1-local diffeomorphism, $\lambda \in (0, 1)$, and let $\Psi_\pm(x)$ be its straightening diffeomorphisms. Then $\Psi_+'(x) < 0$, $\Psi_-'(x) > 0$, and it follows from (2.3.5) that

$$\theta_+(t) = -(\Psi_+^{-1}(t))'e^{-t\ln\lambda}, \quad \theta_-(t) = (\Psi_-^{-1}(t))'e^{-t\ln\lambda}$$

are positive continuous functions satisfying relations

$$\lim_{t\to+\infty} \frac{\theta_\pm(t+1)}{\theta_\pm(t)} = 1. \tag{2.3.9}$$

Hence the representations

$$\Psi_+^{-1}(t) = \int_t^\infty e^{s\ln\lambda}\theta_+(s)ds, \quad \Psi_-^{-1}(t) = -\int_t^\infty e^{s\ln\lambda}\theta_-(s)ds$$

hold.

If, on the other hand, $\theta_\pm(t)$ are arbitrary positive continuous functions satisfying (2.3.9), then the latter representations define a pair of straightening diffeomorphisms $\Psi_\pm(x)$ of some C^1-diffeomorphism $F(x) = \lambda x + o(x)$ defined by (2.3.4).

Summing up the previous discussion we can say that the pair $\theta_\pm(t)$ satisfying (2.3.9) gives a parametrization of C^1-diffeomorphisms $F(x) = \lambda x + o(x)$, $\lambda \in (0,1)$. We will call $\theta_\pm(t)$ the *functional coordinates* of F. If $\tilde\theta_\pm(t)$ is another pair of functional coordinates of the same C^1-diffeomorphism F, then

$$\tilde\theta_\pm(t) = \theta_\pm(H_\pm(t))H_\pm'(t)\, e^{(H_\pm(t)-t)\ln\lambda}$$

with some C^1-diffeomorphisms $H_\pm : \mathbb{R} \to \mathbb{R}$ commuting with the standard shift.

Corollary 2.5. *Let $F(x) = \lambda x + o(x)$ and $\tilde F(x) = \lambda x + o(x)$ with $\lambda \in (0,1)$ be a pair of C^1-diffeomorphisms with functional coordinates $\theta_\pm$ and $\tilde\theta_\pm$, respectively. Then F and $\tilde F$ are locally C^1-conjugate if and only if there exist C^1-diffeomorphisms $H_\pm$ commuting with the standard shift such that*

$$\lim_{t\to+\infty} \frac{\tilde\theta_+(H_+(t))H_+'(t)e^{(H_+(t)-t)\ln\lambda}}{\theta_+(t)} = \lim_{t\to+\infty} \frac{\tilde\theta_-(H_-(t))H_-'(t)e^{(H_-(t)-t)\ln\lambda}}{\theta_-(t)}$$

where both limits are finite and different from zero.

Let, in particular, $\tilde F(x) = \lambda x$ be a linear mapping with $\lambda \in (0,1)$. Following Example 2.6 we choose $\tilde\Psi_+(x) = (\ln\lambda)^{-1}\ln x$ for $x > 0$ and $\tilde\Psi_-(x) = (\ln\lambda)^{-1}\ln|x|$ for $x < 0$. Then $\tilde\Psi_+^{-1}(t) = e^{t\ln\lambda}$, $\tilde\Psi_-^{-1}(t) = -e^{t\ln\lambda}$ and $\tilde\theta_\pm(t) = -\ln\lambda$. According to Corollary 2.5 a diffeomorphism $F(x) = \lambda x + o(x)$ is C^1-conjugate with $\tilde F(x)$ if and only if there exist diffeomorphisms $H_\pm$ commuting with the standard shift such that

$$\lim_{t\to+\infty} \frac{H_+'(t)e^{(H_+(t)-t)\ln\lambda}}{\theta_+(t)} = \lim_{t\to+\infty} \frac{H_-'(t)e^{(H_-(t)-t)\ln\lambda}}{\theta_-(t)} = c \neq 0.$$

Since $H_\pm'(t)\, e^{(H_\pm(t)-t)\ln\lambda}$ are arbitrary positive continuous 1-periodic functions, we arrive at the following proposition.

Corollary 2.6. *A C^1-diffeomorphism F, $F'(0) \in (0,1)$, is C^1-locally conjugate with a linear diffeomorphism if and only if*

$$\theta_{\pm}(t) = p_{\pm}(t)(1 + o(1)), \quad t \to \infty,$$

where $p_{\pm}(t)$ are continuous positive 1-periodic functions.

As we have already seen in Section 2.1, there exist diffeomorphisms hyperbolic at a fixed point which are not smoothly conjugate with a linear one. Now we can construct a continual family of pairwise smoothly non-conjugate hyperbolic diffeomorphisms with the same linear part. Namely, for $\delta \in \mathbb{R}$ let F_δ be a diffeomorphism with $F'_\delta(0) = \lambda \in (0,1)$ and

$$\theta_{\pm}(t) = t^{\delta}, \qquad t > 0.$$

Corollary 2.6 implies that F_δ and F_γ are not C^1-conjugate if $\delta \neq \gamma$.

Let us now consider a non-hyperbolic fixed point. For $\beta \in \mathbb{R}$ and $r \in \mathbb{N}$ we denote by $g_{r,\beta}(x)$ the unique solution of the equation

$$g(x) = r^{-1}\beta x^{-r}\ln(1 + rx^r + \beta x^{2r} + x^{2r}g(x)) - \beta, \tag{2.3.10}$$

analytic in a neighborhood of $x = 0$ and introduce the normal form

$$G_{r,\beta}(x) = \frac{x}{(1 + rx^r + \beta x^{2r} + x^{2r}g_{r,\beta}(x))^{1/r}} \tag{2.3.11}$$

from Corollary 2.3. If we set

$$W_-(x) = \frac{1}{r}x^{-r} + \frac{\beta}{r}\ln x, \qquad x \in \mathbb{R}_-,$$

$$W_+(x) = \frac{1}{r}x^{-r} + \frac{\beta}{r}\ln|x|, \qquad x \in \mathbb{R}_+,$$

then straightforward calculations show that (2.3.10) is equivalent to the relation

$$W_{\pm}G_{r,\beta}(x) = W_{\pm}(x) + 1,$$

and hence $W_{\pm}$ is a pair of straightening diffeomorphisms for $G_{r,\beta}$.

Let F be a C^k-diffeomorphism, $k \geq 2r+2$, such that

$$F(x) = G_{r,\beta}(x) + O(x^{2r+2}).$$

Then $F(x) = H(x) + o(x^k)$ with $H(x) = G_{r,\beta}(x) + c_{2r+2}x^{2r+2} + \cdots + c_k x^k$. According to Corollary 2.3 there exists a formal transformation $\hat{\Phi}$ such that $\hat{\Phi}\hat{H}\hat{\Phi}^{-1} = \hat{G}_{r,\beta}$ where $\hat{H}$ and $\hat{G}_{r,\beta}$ are the formal series of H and $G_{r,\beta}$, respectively. Therefore for every $s \geq 1$ there exists an invertible polynomial transformation Φ_s such that $\Phi_s H \Phi_s^{-1}(x) = G_{r,\beta}(x) + o(x^s)$, and for $s = k$ we obtain

$$\Phi_k F \Phi_k^{-1}(x) = G_{r,\beta}(x) + o(x^k).$$

If $F \in C^\infty$, then the previous relations are fulfilled for every $k \geq 2r+2$. Moreover, for $F \in C^\infty$ there exists an invertible formal series $\hat{\Phi}$ such that $\hat{\Phi}\hat{F}\hat{\Phi}^{-1} = \hat{G}_{r,\beta}$. According to the Borel Theorem there exists a local C^∞-diffeomorphism Φ with the formal series $\hat{\Phi}$ and hence $\mu(x) = \Phi F \Phi^{-1}(x) - G_{r,\beta}(x)$ is a flat function at $x = 0$.

Lemma 2.2. *The straightening diffeomorphisms $\Psi_\pm$ for a local orientation-preserving C^∞-diffeomorphism $F(x) = G_{r,\beta}(x) + \mu(x)$ with a function $\mu(x)$ flat at $x = 0$ may be chosen in the form*

$$\Psi_\pm(x) = W_\pm(x) + \theta_\pm(x)$$

where $\theta_\pm(x)$ are uniquely determined C^∞-functions flat at $x = 0$.

Proof. To simplify notation we define $G(x) \equiv G_{r,\beta}(x)$ and write F in the form

$$F(x) = \frac{G(x)}{(1 + f(x))^{1/r}} \tag{2.3.12}$$

where

$$f(x) = \frac{(G(x))^r - (F(x))^r}{(G(x))^r}$$

is a flat function at $x = 0$. We substitute (2.3.12) in the conjugacy equation

$$\Psi_\pm(F(x)) = \Psi_\pm(x) + 1,$$

take into account that $W_\pm$ satisfy

$$W_\pm(G(x)) = W_\pm(x) + 1$$

and find that $\theta_\pm(x)$ satisfy the cohomological equation

$$\theta_\pm(x) - \theta_\pm(F(x)) = \gamma(x) \tag{2.3.13}$$

where

$$\gamma(x) = \frac{1}{r}(G(x))^{-r} f(x) - \frac{\beta}{r^2}\ln(1 + f(x))$$

is flat at $x = 0$.

Being flat the function $\gamma(x)$ satisfies the estimates

$$|\gamma^{(j)}(x)| \leq a_{jk}|x|^k, \quad k, j = 0, 1, 2, \ldots$$

with some constants $a_{jk} > 0$. In particular,

$$|\gamma(x)| \leq a_{0k}|x|^k, \quad k = 1, 2, \ldots .$$

Besides, as we have already seen in Section 2.2.1, $|F^n(x)| \le cn^{-1/r}$, $x \in [0,\varepsilon]$. Hence the series

$$\theta_+(x) = \sum_{n=0}^{\infty} \gamma(F^n(x))$$

converges uniformly and defines a continuous function θ_+ in $[0,\varepsilon]$, $\theta_+(0) = 0$. It is easy to see that $\theta_+(x)$ is the unique continuous solution of equation (2.3.13) in $[0,\varepsilon]$ vanishing at $x = 0$. Moreover, $\theta_+(x)$ is an infinitely differentiable function in $[0,\varepsilon]$, flat at $x = 0$. To prove this claim it is sufficient to check that (2.3.13) has a C^∞-solution in $[0,\varepsilon]$ which is flat at $x = 0$. To this end let us denote by $R(\{c_{kp}\})$ the convex compact set of C^∞-functions θ in $[0,\varepsilon]$ satisfying the estimates

$$|\theta^{(k)}(x)| \le c_{kp}|x|^p, \quad p,k = 0,1,2,\ldots .$$

Evidently, $\gamma \in R(\{c_{kp}\})$ if $a_{kp} \le c_{kp}$. We will show that with an appropriate choice of an infinite matrix $\{c_{kp}\}$ the set $R(\{c_{kp}\})$ is invariant with respect to the operator

$$L\theta(x) = \theta(F(x)) + \gamma(x).$$

To choose c_{kp} let us note that

$$(L\theta(x))^{(k)} = \theta^{(k)}(F(x))(F'(x))^k + \sum_{j=0}^{k-1} b_{kj}(x)\theta^{(j)}(F(x)) + \gamma^{(k)}(x)$$

where the coefficients $b_{kj}(x)$ are determined by F. Therefore if $\theta \in R(\{c_{kp}\})$, then

$$|(L\theta)(x)| \le c_{0p}|F(x)|^p + a_{0,p+r}|x|^{p+r}, \quad p > 0,$$

and

$$|(L\theta(x))^{(k)}| \le c_{kp}|F(x)|^p|F'(x)|^k + kb_k\tilde{c}_{k-1,p+r}|x|^{p+r} + a_{k,p+r}|x|^{p+r},$$

where

$$b_k = \max_{x\in[0,\varepsilon],0\le j\le k-1} |b_{kj}(x)|, \quad \tilde{c}_{k-1,p+r} = \max_{0\le j\le k-1} c_{j,p+r}.$$

Now we note that according to the representation (2.3.12) and definition of $G(x) = G_{r,\beta}(x)$ by (2.3.11) there exists $\delta > 0$ such that

$$0 \le F(x) \le x(1 - \delta x^r), \quad 0 < F'(x) \le 1, \quad x \in [0,\varepsilon].$$

Choose c_{0p} such that

$$(1 - \delta x^r)^p + \frac{a_{0,p+r}}{c_{0p}} x^r \le 1, \quad x \in [0,\varepsilon],\ p > 0,$$

and then choose recursively c_{kp} such that

$$(1 - \delta x^r)^p + \frac{kb_k\tilde{c}_{k-1,p+r} + a_{k,p+r}}{c_{kp}} x^r \le 1.$$

With such a choice of c_{kp} the set $R(\{c_{kp}\})$ is invariant with respect to the operator L and by the Leret–Shauder theorem [42] there exists a solution $\theta \in R(\{c_{kp}\})$ of (2.3.13). Since the continuous solution of (2.3.13) in $[0,\varepsilon]$ is unique, we obtain $\theta(x) = \theta_+(x)$ implying $\theta_+ \in C^\infty[0,\varepsilon]$.

If r is even we use the same arguments to find that the series

$$\theta_-(x) = \sum_{n=0}^{\infty} \gamma(F^n(x)), \quad x \in [-\varepsilon, 0],$$

defines a C^∞-solution of (2.3.13). If r is odd, then

$$0 \le |F^{-1}(x)| \le |x|(1 + \delta x^r), \ \ 0 < (F^{-1}(x))' \le 1, \ \ x \in [-\varepsilon, 0],$$

and the series

$$\theta_-(x) = -\sum_{n=1}^{\infty} \gamma(F^{-n}(x))$$

defines a solution of (2.3.13) in $[-\varepsilon, 0]$. □

The following statement shows that the functions $\theta_\pm(x)$ appearing in Lemma 2.2 are not essential for the C^∞-classification of C^∞-diffeomorphisms $F(x) = G_{r,\beta}(x) +$ flat residual.

Corollary 2.7. *Every local C^∞- diffeomorphism $F(x) = G_{r,\beta}(x) + \mu(x)$ with a flat residual $\mu(x)$ is C^∞-conjugate with $G_{r,\beta}(x)$.*

Proof. We accept $H_\pm(x) = x$ in (2.3.7) and set

$$\Phi(x) = \begin{cases} \Psi_+^{-1}(W_+(x)), & x > 0, \\ 0, & x = 0, \\ \Psi_-^{-1}(W_-(x)), & x < 0, \end{cases}$$

where $\Psi_\pm$ are straightening diffeomorphisms for F. According to Lemma 2.2 the differences $\theta_\pm(x) = \Psi_\pm(x) - W_\pm(x)$ are flat at the origin and hence Φ is a local C^∞-diffeomorphism of the form $\Phi(x) = x +$ flat residual satisfying

$$\Psi_\pm(\Phi(x)) = W_\pm(x).$$

Theorem 2.8 implies that F and $G_{r,\beta}$ and C^∞-conjugate. □

Proof of Theorem 2.7 for local C^∞-diffeomorphisms. Let F and $\tilde{F}$ be local C^∞-diffeomorphisms at the joint fixed point $x = 0$, and $\hat{F}^2(x) \ne x$. Assume that F and $\tilde{F}$ are formally conjugate and prove that they are locally C^∞-conjugate as well.

Diffeomorpisms with hyperbolic fixed points were treated in Theorem 2.5 and hence it is sufficient to consider the case $|F'(0)| = 1$. If F preserves the orientation, then $F'(0) = 1$ and according to Corollary 2.3 we can assume that

$F(x) = G_{r,\beta}(x) +$ flat residual. Now Corollary 2.7 states that both F and $\tilde{F}$ are conjugate with $G_{r,\beta}$.

Let now $F'(0) = \tilde{F}'(0) = -1$. Since F and $\tilde{F}$ are formally conjugate, we can assume that $F(x) - \tilde{F}(x)$ and hence $F^2(x) - \tilde{F}^2(x)$ are flat at $x = 0$. Now $F^2(x)$ and $\tilde{F}^2(x)$ preserve the orientation and, as we have just proved, are conjugate: $\tilde{F}^2(x) = \Phi F^2 \Phi^{-1}(x)$ where $\Phi(x) = x +$ flat residual. If

$$F_1 = \Phi F \Phi^{-1},$$

then $F_1(x) = F(x) +$ flat residual and the local C^∞-diffeomorphism

$$\Psi(x) = \begin{cases} x, & x \le 0, \\ F_1 \tilde{F}^{-1}(x), & x \ge 0, \end{cases}$$

has the form $\Psi(x) = x +$ flat residual. Besides, $F_1^2 = \Phi F^2 \Phi^{-1} = \tilde{F}^2$. If $x \le 0$, then $\Psi \tilde{F} \Psi^{-1}(x) = \Psi \tilde{F}(x) = F_1 \tilde{F}^{-1} \tilde{F}(x) = F_1(x)$. For $x \ge 0$ we have $\Psi \tilde{F} \Psi^{-1}(x) = \Psi \tilde{F} \tilde{F} F_1^{-1}(x) = \tilde{F}^2 F_1^{-1}(x) = F_1(x)$. It means that $\tilde{F}$ and F_1 are conjugate, which implies that $\tilde{F}$ and F are conjugate as well. □

2.3.3 Analytic classification

If a local orientation-preserving diffeomorphism $F : \mathbb{R} \to \mathbb{R}$ is analytic at $x = 0$, then for every finite k there exists a polynomial transformation Φ such that

$$\Phi F \Phi^{-1}(x) = G_{r,\beta}(x) + O(x^k).$$

Hence we can assume that

$$F(x) = G_{r,\beta}(x) + O(x^k), \quad k \ge 2r + 2.$$

Our original intention was to consider a diffeomorphism F at a neighborhood of $x = 0$ on the *real* line. As we have already seen, a smooth classification of smooth diffeomorphisms is based on the existence of two straightening diffeomorphisms defined on semi-neighborhoods of $x = 0$. A local real-analytic diffeomorphism F is uniquely continued to a *complex* neighborhood $D(\varepsilon) = \{ z : |z| < \varepsilon \}$ of $x = 0$ and as a result some additional straightening diffeomorphisms arise. To describe them let us fix $\alpha \in (\pi(2r)^{-1}, \pi r^{-1})$ and consider the sectors

$$S_j = \{ z : z \ne 0, \ |\arg z - \frac{j\pi}{r}| < \alpha \}, \quad j = 1, \dots, 2r.$$

It is evident that these sectors form an open covering of the punctured plane $\mathbb{C} \setminus \{0\}$. We denote by $U_j(\varepsilon), j = 1, \dots, 2r$, the intersection of S_j with a complex neighborhood $D(\varepsilon) = \{ z : |z| < \varepsilon \}$ of $z = 0$.

Lemma 2.3. *There exists sufficiently small* $\varepsilon > 0$ *such that for every integer* j, $1 \le j \le 2r$, *there exists the unique function* θ_j *analytic in* $U_j = U_j(\varepsilon)$, *continuous in its closure* $\overline{U}_j$ *and vanishing at zero, such that for any choice of a continuous branch of* $\ln z$ *in* S_j *and of a constant* $d_j \in \mathbb{C}$ *the diffeomorphism*

$$\Psi_j(z) = \frac{1}{r} z^{-r} + \frac{\beta}{r} \ln z + d_j + \theta_j(z), \quad z \in U_j, \tag{2.3.14}$$

conjugates $F|_{U_j}$ *with the standard shift if* j *is even, and* $F^{-1}|_{U_j}$ *with the inverse standard shift if* j *is odd.*

Proof. We start with a small $\varepsilon = \varepsilon_0 > 0$ such that F is analytic and one-to-one in $D(\varepsilon_0)$. The function $\omega(z) = r^{-1}z^{-r}$, which is a "principal part" of $\Psi_j(z)$ from (2.3.14), maps $U_j(\varepsilon_0)$ biholomorphically on the domain

$$W_j(\varepsilon_0) = \{ s : |s| > r^{-1}\varepsilon_0^{-1}, \ |\arg s + j\pi| < \alpha r \},$$

and if we set $G_j(s) = \omega G_{r,0} \omega^{-1}(s)$, $s \in W_j(\varepsilon_0)$, where $G_{r,0}(z) \equiv z(1 + rz^r)^{-1/r}$ is a "principal part" of F, then

$$G_j(s) = s + 1, \quad s \in W_j(\varepsilon_0), \ j = 1, \dots, 2r.$$

Now we define the diffeomorphisms $F_j = \omega F \omega^{-1}$ in $W_j(\varepsilon_0)$, $j = 1, \dots, 2r$, and after straightforward calculations find

$$F_j(s) = s + 1 + O(s^{-1}), \quad s \in W_j(\varepsilon_0), \ j = 1, \dots, 2r. \tag{2.3.15}$$

Therefore for every even j there exists $\varepsilon \in (0, \varepsilon_0)$ such that for each $s \in \overline{W_j(\varepsilon)}$ all points $F_j^n(s)$, $n = 0, 1, 2, \dots$, belong to $\overline{W_j(\varepsilon_0)}$. Similarly, for every odd j and each $s \in \overline{W_j(\varepsilon)} \subset \overline{W_j(\varepsilon_0)}$ all points $F_j^{-n}(s)$, $n = 1, 2, \dots$, belong to $\overline{W_j(\varepsilon_0)}$. Moreover, the asymptotic representation (2.3.15) implies that for every compact set $K \subset \overline{W_j(\varepsilon)}$ there exists a constant $c > 0$ such that

$$\left|F_{2q}^n(s)\right| \ge cn, \quad \left|F_{2q-1}^{-n}(s)\right| \ge cn, \quad s \in K, \ n = 1, 2, \dots .$$

From now on we fix $\varepsilon > 0$ and $U_j = U_j(\varepsilon)$ and, similar to the proof of Lemma 2.2, set

$$G(z) = G_{r,\beta}(z), \quad f(z) = \frac{(G(z))^r - (F(z))^r}{(F(z))^r}, \quad z \in D(\varepsilon).$$

If (2.3.14) defines a diffeomorphism Ψ_j conjugating $F|_{U_j}$ with the standard shift, then the function $\theta_j(z)$ must satisfy the cohomological equation

$$\theta_j(z) - \theta_j(F(z)) = \gamma(z), \quad z \in U_j, \ j = 1, \dots, 2r, \tag{2.3.16}$$

where

$$\gamma(z) = \frac{1}{r}(G(z))^{-r} f(z) - \frac{\beta}{r^2} \ln(1 + f(z)) \tag{2.3.17}$$

and the branch of $\ln(1 + f(z))$ is such that $\ln 1 = 0$.

It is evident that $\gamma(z) = O(z^{k-r-1})$. If we now set

$$\sigma_j(s) = \theta_j(\omega^{-1}(s)), \quad \gamma_j(s) = \gamma(\omega^{-1}(s)), \quad s \in W_j(\varepsilon), \quad j = 1, \ldots, 2r,$$

then we obtain the cohomological equation

$$\sigma_j(s) - \sigma_j(F_j(s)) = \gamma_j(s), \quad s \in \overline{W_j(\varepsilon)}, \ j = 1, \ldots 2r, \tag{2.3.18}$$

equivalent to (2.3.16). The functions $\gamma_j(s)$ satisfy the estimates

$$|\gamma_j(s)| \le c|\omega^{-1}(s)|^{k-r-1} = c|s|^{-\frac{k-r-1}{r}}, \quad s \in \overline{W_j(\varepsilon)}, \ j = 1, \ldots, 2r,$$

and hence

$$\left|\gamma_{2q}(F_{2q}^n(s))\right| \le c(K)n^{-\frac{k-r-1}{r}}, \quad \left|\gamma_{2q-1}(F_{2q-1}^{-n}(s))\right| \le c(K)n^{-\frac{k-r-1}{r}},$$

$$s \in \overline{W_j(\varepsilon)} \bigcap K, \ j = 1, \ldots, 2r,$$

for every compact set $K \subset \mathbb{C}$.

Since $k \ge 2r + 2$, the series

$$\sigma_{2q}(s) = \sum_{n=0}^{\infty} \gamma_{2q}(F_{2q}^n(s)), \quad s \in W_{2q}(\varepsilon), \ q = 1, \ldots, r,$$

and

$$\sigma_{2q-1}(s) = -\sum_{n=1}^{\infty} \gamma_{2q-1}(F_{2q-1}^{-n}(s)), \quad s \in W_{2q-1}(\varepsilon), \ q = 1, \ldots, r,$$

converge in corresponding domains $\overline{W_j(\varepsilon)}$ and define solutions $\sigma_j(s)$ of (2.3.18) analytic in $W_j(\varepsilon)$. To complete the proof we come back to the functions $\theta_j(z) = \sigma_j(\omega(z))$ and obtain solutions of (2.3.16) analytic in U_j, $j = 1, \ldots 2r$, and representable by series

$$\theta_{2q}(z) = \sum_{n=0}^{\infty} \gamma(F^n(z)), \ z \in U_{2q}; \ \theta_{2q-1}(z) = -\sum_{n=1}^{\infty} \gamma(F^{-n}(z)), \ z \in U_{2q-1};$$

$$q = 1, \ldots, r. \tag{2.3.19}$$

□

Similar to the smooth situation, the collection $\{ \theta_j \}_{j=1}^{2r}$ may be considered as some functional coordinates of F. Now we fix a branch of $\ln z$, $z \in S_j$ for every $j = 1, 2, \ldots 2r$ and choose $d_j = 0$. Then the straightening diffeomorphisms Ψ_j depend on F only.

Theorem 2.9. *Analytic diffeomorphisms*

$$F(x) = G_{r,\beta}(x) + O(x^k), \quad \tilde{F}(x) = G_{r,\beta}(x) + O(x^k)$$

with $k \geq 2r + 2$ *and functional coordinates* $\{\theta_j\}_{j=1}^{2r}, \{\tilde{\theta}_j\}_{j=1}^{2r}$, *respectively, are locally analytically conjugate if and only if there exists a locally analytic function* $\Phi(z) = z + dz^{r+1} + O(z^{r+2})$ *such that*

$$\tilde{\theta}_j(\Phi(z)) = \theta_j(z) + \frac{1}{r}(z^{-r} - \Phi^{-r}(z)) + \frac{\beta}{r}\ln\frac{z}{\Phi(z)} - d,$$

$$z \in U_j, \qquad j = 1, \ldots, 2r. \tag{2.3.20}$$

Proof. Let $\tilde{F} = \Phi F \Phi^{-1}$. Then for every fixed j, $j = 1, \ldots, 2r$, we have

$$\tilde{\Psi}_j^{-1} T \tilde{\Psi}_j(z) = \Phi \Psi_j^{-1} T \Psi_j \Phi^{-1}(z), \quad z \in S_j,$$

where Ψ_j are the straightening diffeomorphisms for F and $\tilde{F}$, respectively, from Lemma 2.3, and T is the standard shift. Hence the mappings $\tilde{\Psi}_j \Phi \Psi_j^{-1}$ commute with T and

$$\tilde{\Psi}_j \Phi \Psi_j^{-1}(t) = t + \tau_j(t), \quad t \in \Psi_j(U_j),$$

where τ_j are 1-periodic entire functions. Therefore the expression

$$\tilde{\Psi}_j(\Phi(z)) - \Psi_j(z) \equiv \tau_j(\Psi_j(z)), \ z \in U_j(\varepsilon),$$

is bounded on the set

$$\{ z : z = \Psi_j^{-1}(t), \ t \in \mathbb{R} \bigcap \Psi_j(U_j) \}.$$

The boundedness together with (2.3.14) implies that there exists a constant d such that a representation

$$\Phi(z) = z + dz^{r+1} + O(z^{r+2}), \ z \in D(\varepsilon),$$

is valid in a neighborhood of $z = 0$. Using again (2.3.14) we find

$$\lim_{z \to 0} (\tilde{\Psi}_j(\Phi(z)) - \Psi_j(z)) = -d,$$

and hence

$$\lim_{t \to \infty, \, t \in \mathbb{R}} \tau_j(t) = -d.$$

Being 1-periodic entire function, $\tau_j(t)$ is identically constant, and hence

$$\tilde{\Psi}_j(\Phi(z)) = \Psi_j(z) - d, \ z \in U_j,$$

which coincides with (2.3.20).

Conversely, if (2.3.20) holds, then the latter identities follow and the mappings

$$\tilde{\Psi}_j \Phi \Psi_j^{-1}(t) = t - d, \quad t \in \Psi_j(U_j),$$

commute with the standard shift:

$$T\tilde{\Psi}_j \Phi \Psi_j^{-1} = \tilde{\Psi}_j \Phi \Psi_j^{-1} T.$$

This equality is equivalent to

$$\tilde{\Psi}_j^{-1} T \tilde{\Psi}_j(z) = \Phi \Psi_j^{-1} T \Psi_j \Phi^{-1}(z), \quad z \in U_j,$$

which implies

$$\tilde{F} = \Phi F \Phi^{-1},$$

completing the proof of Theorem 2.9. □

Let us clarify conditions for $F(x) = G_{r,\beta}(x) + O(x^{2r+2})$ to be conjugate with $G_{r,\beta}(x)$. If we set $\tilde{F} = G_{r,\beta}$ in Theorem 2.9, then we obtain $\tilde{\theta}_j(z) = 0$, $z \in U_j$, $j = 1, \dots, 2r$, and as (2.3.20) shows every function $\theta_j(z)$, $z \in U_j$, coincides with the function

$$h(z) = \frac{1}{r}(\Phi^{-r}(z) - z^{-r}) + \frac{\beta}{r} \ln \frac{\Phi(z)}{z} + d, \tag{2.3.21}$$

analytic in a non-punctured neighborhood of $z = 0$. Hence

$$\theta_{2q}(z) - \theta_{2q-1}(z) = 0, \quad z \in U_{2q} \bigcap U_{2q-1}, \; q = 1, \dots, r, \tag{2.3.22}$$

and (2.3.19) yields

$$\sum_{n=-\infty}^{\infty} \gamma(F^n(z)) = 0, \quad z \in U_{2q} \bigcap U_{2q-1}, \quad q = 1, \dots, r, \tag{2.3.23}$$

where $\gamma(z)$ is defined by (2.3.17). On the other hand, if (2.3.23) holds, then according to (2.3.19) the coordinate functions $\theta_j(z)$, $j = 1, \dots, 2r$, satisfy (2.3.22) and

$$h(z) = \theta_j(z), \quad z \in U_j, \quad j = 1. \dots, 2r,$$

is a single-valued analytic function at a neighborhood of $z = 0$. Starting with $h(z)$ we can find the unique single-valued function $\Phi(z)$ analytic at $z = 0$ such that $\Phi(z) = z + dz^{r+1} + O(z^{r+2})$ and (2.3.21) is fulfilled. It means that local parameters $\{\tilde{\theta}_j = 0\}_{j=1}^{2r}$ for $G_{r,\beta}$ and $\{\theta_j\}_{j=1}^{2r}$ for F satisfy conditions of Theorem 2.9 and the following statement is true.

Corollary 2.8. *A local analytic diffeomorphism* $F(x) = G_{r,\beta}(x) + O(x^k)$, $k \geq 2r + 2$, *is analytically conjugate with* $G_{r,\beta}(x)$ *if and only if* (2.3.23) *holds.*

Example 2.8. The local diffeomorphisms

$$F_\delta(x) = \frac{x}{1 + x + \delta x^3}, \quad \delta \in \mathbb{R},$$

and

$$G_{1,0}(x) = \frac{x}{1 + x}$$

have the same formal normal form $x - x^2 + x^3$ from Theorem 2.6. Let us show that they are not analytically conjugate, at least for sufficiently small $\delta \neq 0$. To this end we note that since here $r = 1$, the covering $\{S_j\}$ contains two sectors only, namely,

$$S_1 = \{ z \neq 0, |\arg z - \pi| < \alpha\}, \quad S_2 = \{ z \neq 0, |\arg z - 2\pi| < \alpha\},$$

with $\alpha \in (\frac{\pi}{2}, \pi)$. Since $\gamma(z) = \delta z^2$, and $it \in S_1 \bigcap S_2$, $t \in \mathbb{R}$, it is sufficient to check that

$$\sum_{n=-\infty}^{\infty} (F_\delta^n(it))^2 \neq 0,$$

for $t \in \mathbb{R}$ and δ sufficiently small. It is easy to check that an estimate

$$| F_\delta^n(it)| \leq \frac{c}{|n|}$$

is fulfilled with a constant c for all $t \in \mathbb{R}$ and all sufficiently small $\delta > 0$. Therefore the series

$$\rho(t, \delta) = \sum_{n=-\infty}^{\infty} (F_\delta^n(it))^2$$

converges uniformly with respect to (t, δ) in a neighborhood of $0 \in \mathbb{R}^2$, and defines a continuous function. For $\delta = 0$ we obtain

$$\rho(t, 0) = \sum_{n=-\infty}^{\infty} \frac{1}{(it^{-1} + n)^2} = - \frac{\pi^2}{\sinh^2(\pi t^{-1})} \neq 0,$$

and hence the function $\rho(t, \delta)$ does not vanish if δ and t are sufficiently small.

The same arguments show that a local diffeomorphism

$$F(x) = \frac{x}{(1 + rx^r + f(x))^{1/r}}$$

where $f(x) = \delta_{2r+1} x^{2r+1} + \cdots + \delta_k x^k$ is a polynomial, $f \neq 0$, with sufficiently small coefficients $\delta_{2r+1}, \ldots, \delta_k$ is not analytically conjugate with $G_{r,0}(x)$.

The coordinate functions $\{\theta_j\}_{j=1}^{2r}$ appearing in Theorem 2.9 cannot be arbitrary. Namely, since the diffeomorphisms

$$\Psi_{j+1}\Psi_j^{-1} : \; \Psi_j(U_j \bigcap U_{j+1}) \to \Psi_{j+1}(U_j \bigcap U_{j+1}), \; j = 1, \ldots, 2r;$$

$$U_{2r+1} = U_1; \quad \Psi_{2r+1} = \Psi_1,$$

commute in respective domains with the standard shift, the functions

$$h_j(t) = \Psi_{j+1}\Psi_j^{-1}(t) - t, \quad t \in U_j \bigcap U_{j+1}; \quad j = 1, \dots, 2r,$$

are extendable to a non-connected set $\{ t : |\Im t| > \varepsilon^{-1} \}$ as analytic 1-periodic functions. On the other hand, the collection $\{ \Psi_j, U_j \}$, may be considered as an atlas of the analytic structure of a punctured neighborhood of the origin with the transition functions $\Psi_{j+1}\Psi_j^{-1}(t) = t + h_j(t)$. The following proposition states that up to a simple equivalence this collection forms a complete system of invariants with respect to the analytic conjugacy.

Corollary 2.9. *Let*

$$F(x) = G_{r,\beta}(x) + O(x^k), \quad \tilde{F}(x) = G_{r,\beta}(x) + O(x^k), \quad k \geq 2r + 2,$$

be two local analytic diffeomorphisms with collections of 1*-periodic analytic functions* $\{h_j\}$ *and* $\{\tilde{h}_j\}$*, respectively. Then* F *and* $\tilde{F}$ *are analytically conjugate if and only if* $\tilde{h}_j(t) = h_j(t + d)$, $j = 1, \dots, 2r$, *with a constant* $d \in \mathbb{C}$.

Proof. If $\tilde{F} = \Phi F \Phi^{-1}$, then (2.3.20) is fulfilled which is equivalent to

$$\tilde{\Psi}_j(\Phi(z)) = \Psi_j(z) + d, \quad z \in U_j, \quad j = 1, \dots, 2r.$$

Therefore

$$\tilde{\Psi}_{j+1}\tilde{\Psi}_j^{-1}(t + d) = \Psi_{j+1}\Psi_j^{-1}(t) + d, \quad j = 1, \dots, 2r,$$

i.e.,

$$\tilde{h}_j(t + d) = h_j(t), \quad j = 1, \dots, 2r.$$

Conversely, let the previous equality hold. Then the function

$$\Phi(z) = \tilde{\Psi}_j^{-1} H \Psi_j(z), \quad z \in U_j, \quad j = 1, \dots, r,$$

where $H(t) = t + d$, is a well-defined local analytic diffeomorphism conjugating F and $\tilde{F}$. □

Collections $\{h_j\}_{j=1}^{2r}$ of analytic 1-periodic functions up to the above equivalence are called *Ecalle–Voronin moduli.* More detailed description of these moduli including a realization theorem may be found in [65].

2.4 Diffeomorphisms with isolated fixed points

Diffeomorphisms from the title form a class (one from not many !) of mappings for which it is possible to give a sufficiently complete classification, both topological and smooth.

We suppose here that $F : M \to M$ is either a homeo- or diffeomorphism with $\mathrm{Fix}\, F \neq \emptyset$ and such that all its non-wandering points are fixed. If $M = \mathbb{R}$, then, in particular, the presence of fixed points implies that either F increases (i.e., preserves the orientation) or F has exactly one fixed point and has no periodic point of period 2. If $M = \mathbb{T}$, then F either preserves orientation or F has exactly two fixed points and F permutes the corresponding arcs. To be definite, we consider mappings which preserve the orientation.

2.4.1 Topological classification

Let $U = (x', x'')$ be the interval adjacent to the set $\mathrm{Fix}\, F$ ordered by the orientation of M. Then U is F-invariant and there may be two types of dynamical behavior of F on U: either $F^n(x) \to x''$ or $F^{-n}(x) \to x''$ as $n \to +\infty$.

Theorem 2.10. *Let F and $\tilde{F}$ be two orientation-preserving homeomorphisms with the same set of fixed points and the same dynamical behavior on every interval adjacent to $\mathrm{Fix}\, F = \mathrm{Fix}\, \tilde{F}$. Then F and $\tilde{F}$ are topologically conjugate.*

Proof. Let $U = (x', x'')$ be an interval adjacent to $\mathrm{Fix}\, F = \mathrm{Fix}\, \tilde{F}$. Since the restrictions $F|_U$ and $\tilde{F}|_U$ are fixed-point free homeomorphisms, then according to Theorem 2.1 there exists a homeomorphism $\Phi_U : U \to U$ conjugating these restrictions. Since the dynamical behavior of F and $\tilde{F}$ on U is the same, Φ_U preserves the orientation on M which means

$$\Phi_U(x) \to x', \quad x \to x'; \quad \Phi_U(x) \to x'', \quad x \to x''.$$

With Φ_U being chosen on every adjacent interval we set

$$\Phi(x) = \begin{cases} \Phi_U(x), & x \in U, \\ x, & x \in \mathrm{Fix}\, F = \mathrm{Fix}\, \tilde{F}, \end{cases}$$

and obtain a homeomorphism $\Phi : M \to M$ conjugating F and $\tilde{F}$. □

Let us note that the above proof is valid for mappings F and $\tilde{F}$ with an arbitrary set $\mathrm{Fix}\, F = \mathrm{Fix}\, \tilde{F}$.

2.4.2 Smooth classification of diffeomorphisms with a unique fixed point

First of all we will show that the local conjugacy class of a diffeomorphism $F : \mathbb{R} \to \mathbb{R}$ with a unique fixed point completely defines its global conjugacy class.

Theorem 2.11. *Let F and $\tilde{F}$ be two C^k-diffeomorphisms of the real line with the unique fixed point x_0. If F and $\tilde{F}$ are locally C^k-conjugate at x_0, then they are C^k-conjugate on the entire real line.*

Proof. Let x_0 be the unique fixed point of F and $\tilde{F}$, and let Φ_0 be a local C^k-diffeomorphism conjugating F and $\tilde{F}$ at a neighborhood U of x_0, i.e.,

$$\Phi_0 F(x) = \tilde{F}\Phi_0(x), \quad x \in U.$$

We extend Φ_0 to the entire real axis using relations

$$\Phi(x) = \begin{cases} \tilde{F}^{-n}\Phi_0 F^n(x), & F^n(x) \in U, \quad x_0 \text{ is attracting}, \\ \tilde{F}^{n}\Phi_0 F^{-n}(x), & F^{-n}(x) \in U, \quad x_0 \text{ is repelling}. \end{cases}$$

If x_0 is a point of overflow from the left to the right, then the extension is given by the relations

$$\Phi(x) = \begin{cases} \tilde{F}^{n}\Phi_0 F^{-n}(x), & x \geq x_0, \ F^{-n}(x) \in U, \\ \tilde{F}^{-n}\Phi_0 F^{n}(x), & x \leq x_0, \ F^{n}(x) \in U. \end{cases}$$

□

Let us now consider the case $M = \mathbb{T}$. As we will show the situation here is different from that of the real line: the local conjugacy of diffeomorphisms F and $\tilde{F}$ at the unique fixed point does not imply their global smooth conjugacy.

Denote by $C(k, G)$ the set of all C^k-diffeomorphisms of the circle with a unique fixed point and locally C^k-conjugate at this point with a fixed C^k-mapping G defined in some neighborhood of $x_0 = 0$ on the real line, $G(0) = 0$. It is evident that $x = 0$ must necessarily be a point of overflow for G.

According to the definition, for every diffeomorphism $F \in C(k, G)$ with a fixed point z_0 there exist a neighborhood U_0 of z_0 and a local C^k-diffeomorphism $\Psi_0 : U_0 \to V$, $\Psi_0(z_0) = 0, V \subset \mathbb{R}$, such that

$$F|_{U_0} = \Psi_0^{-1} G \Psi_0.$$

If L is an analytic diffeomorphism mapping $\mathbb{T}\setminus\{z_0\}$ onto $\mathbb{R}$, then $LFL^{-1} : \mathbb{R} \to \mathbb{R}$ is a C^k-diffeomorphism without fixed points. According to Corollary 2.2 it is C^k-conjugate with the standard shift T. Finally,

$$F|_{\mathbb{T}\setminus z_0} = \Psi^{-1} T \Psi$$

where $\Psi : \mathbb{T}\setminus\{z_0\} \to \mathbb{R}$ is a C^k-diffeomorphism.

The intersection $W = U_0 \bigcap \mathbb{T} \setminus \{z_0\}$ is a punctured neighborhood of z_0, and

$$\Psi_0^{-1} G \Psi_0|_W = \Psi^{-1} T \Psi|_W.$$

Hence the mapping

$$D(t) = \Psi_0 \Psi^{-1}(t), \quad t \in \Psi(W),$$

is a C^k-diffeomorphism from a neighborhood of a point at infinity onto a punctured neighborhood of $0 \in \mathbb{R}$, satisfying the equation

$$GD(t) = D(t+1), \quad |t| > N, \tag{2.4.1}$$

with some $N > 0$. A neighborhood of a point at infinity consists of two infinite intervals $V_- = (-\infty, -N)$ and $V_+ = (N, +\infty)$. Hence D is formed by a pair of C^k-diffeomorphisms D_+ and D_- defined on V_+ and V_-, respectively, and mapping these rays onto respective semi-neighborhoods of the origin.

The pair $\{D_+, D_-\}$ is not uniquely determined by the diffeomorphism F. It is easy to see that if S and H are C^k-diffeomorphisms commuting with G and the standard shift, respectively, then the pair

$$\tilde{D}_\pm = S^{-1} D_\pm H \tag{2.4.2}$$

corresponds to the same diffeomorphism F.

Relation (2.4.2) determines an equivalence of C^k-pairs $D_\pm$ satisfying (2.4.1) and if we denote by $J(k, G)$ the set of all equivalence classes, then the correspondence $F \to D_\pm$ defines a mapping $I : C(k, G) \to J(k, G)$.

Theorem 2.12. *Two diffeomorphisms F and $\tilde{F}$ belonging to $C(k, G)$ are C^k-conjugate if and only if $I(F) = I(\tilde{F})$.*

The mapping I is surjective: for every $c \in J(k, G)$ there exists a diffeomorphism $F \in C(k, G)$ such that $I(F) = c$.

Proof. Let $\tilde{F} = \Phi F \Phi^{-1}$, $\Phi \in C^k$. Then

$$\tilde{\Psi}_0^{-1} G \tilde{\Psi}_0|_{U_0} = \Phi \Psi_0^{-1} G \Psi_0 \Phi^{-1}|_{U_0}$$

and

$$\tilde{\Psi}^{-1} T \tilde{\Psi}|_{\mathbb{T}\setminus\{z_0\}} = \Phi \Psi^{-1} T \Psi \Phi^{-1}|_{\mathbb{T}\setminus\{z_0\}},$$

where $\tilde{\Psi}_0$, $\tilde{\Psi}$ (and Ψ_0, Ψ) are C^k-diffeomorphisms reducing $\tilde{F}$ (and F) to G and T, respectively. It follows that the local C^k-diffeomorphisms

$$S = \tilde{\Psi}_0 \Phi \Psi_0^{-1}, \quad H = \tilde{\Psi} \Phi \Psi^{-1},$$

commute with G and with the standard shift, respectively. Since

$$\Phi|_{U_0} = \tilde{\Psi}_0^{-1} S \Psi_0, \quad \Phi|_{\mathbb{T}\setminus\{z_0\}} = \tilde{\Psi}^{-1} H \Psi,$$

we arrive at the identity

$$D_{\pm}(t) = S^{-1}\tilde{D}_{\pm}H(t) \tag{2.4.3}$$

proving $I(F) = I(\tilde{F})$.

Conversely, let $I(F) = I(\tilde{F})$. Then (2.4.3) holds with some C^k-diffeomorphisms S and H commuting with G and T, respectively. The mapping

$$\Phi(x) = \begin{cases} \tilde{\Psi}_0^{-1}S\Psi_0(x), & x \in U_0, \\ \tilde{\Psi}^{-1}H\Psi(x), & x \in \mathbb{T}\setminus\{z_0\}, \end{cases}$$

is a well-defined C^k-diffeomorphism of $\mathbb{T}$. It is easy to check that Φ conjugates F and $\tilde{F}$.

To prove the remaining part of Theorem 2.12, let $D = D_{\pm}(t)$ be a representative of a class $c \in J(k, G)$. To find an appropriate diffeomorphism F, let us consider D as a diffeomorphism of a neighborhood of a point at infinity onto a punctured neighborhood of the origin $0 \in \mathbb{R}$. If

$$\Psi : \mathbb{T}\setminus\{z_0\} \to \mathbb{R}$$

is a fixed homeomorphism, $\Psi(z) \to \infty$ as $z \to z_0$, then

$$\lim_{z\to z_0} D\Psi(z) = 0$$

and

$$\Psi_0(z) \equiv D\Psi(z)$$

is a local homeomorphism of a neighborhood $U_0 \subset \mathbb{T}$ in a neighborhood of $0 \in \mathbb{R}$.

The maps (U_0, Ψ_0) and $(\mathbb{T}\setminus\{z_0\}, \Psi)$ form an atlas of a C^k-manifold Z homeomorphic to $\mathbb{T}$ with the transition function $D = \Psi_0\Psi^{-1}$. It is well known [69] that there exists a unique smooth structure on $\mathbb{T}$ and therefore there exists a homeomorphism $W : Z \to \mathbb{T}$ such that $W\Psi_0^{-1}$ and $W\Psi^{-1}$ are C^k-functions.

Since $D = \Psi_0 W^{-1}(\Psi W^{-1})^{-1}$, we can assume that Ψ_0 and Ψ are C^k-mappings. Now we set

$$F(x) = \begin{cases} \Psi_0^{-1}G\Psi_0(x), & x \in U_0, \\ \Psi^{-1}T\Psi(x), & x \in \mathbb{T}\setminus\{z_0\}, \end{cases}$$

and by virtue of (2.4.1) obtain a well-defined C^k-diffeomorphism of $\mathbb{T}$ belonging to $C(k, G)$. Since $\Psi^{-1}\Psi_0 = D$, we conclude $I(F) = c$. □

The "moduli manifold" $J(k, G)$ may be described explicitly for some specific choice of G. Let, for example,

$$G(x) = G_{r,0}(x) = \frac{x}{(1 + rx^r)^{1/r}}.$$

Since $x = 0$ is a point of overflow, then r is odd, and a diffeomorphism $D = D_\pm$ satisfying (2.4.1) has the form

$$D_\pm(t) = (rt + d_\pm(t))^{-\frac{1}{r}}$$

where $d_\pm(t)$ are 1-periodic C^k-functions satisfying

$$d'_\pm(t) + r \neq 0, \quad t \in \mathbb{R}.$$

Let $S(x) = \lambda x + o(x)$, $\lambda \neq 0$, be a local C^1-diffeomorphism commuting with $G_{r,0}$. Then

$$\frac{S(x)}{(1 + rS^r(x))^{1/r}} = S\left(\frac{x}{(1 + rx^r)^{1/r}}\right)$$

and the function $L(t) = r^{-1}S^{-r}(t^{-1/r})$ satisfies the equation $L(t+r) = L(t)+1$ in a neighborhood of the point at infinity. Hence $L(t) = r^{-1}(t + \mu(t))$ with some r-periodic function μ and $S(x) = (x^{-r} + \mu(x^{-r}))^{-1/r}$. Differentiating we find

$$\lambda = \lim_{x\to 0} S'(x) = \lim_{x\to 0}(1 + \mu'(x^{-r}) + o(1))$$

which is possible if and only if $\mu(x) \equiv \text{const}$, $\lambda = 1$. Therefore every local C^1-diffeomorphism S commuting with $G_{r,0}$ has the form

$$S(x) = \frac{x}{(1 + \mu x^r)^{1/r}}$$

where μ is a real constant. With $W_\pm(t) = D_\pm^{-r}(t)$, $\tilde{W}_\pm(t) = \tilde{D}_\pm^{-r}(t)$, relations (2.4.1) yield $W_\pm(t + 1) = W_\pm(t) + r$, $\tilde{W}_\pm(t + 1) = \tilde{W}_\pm(t) + r$, and

$$W_\pm(t) = rt + d_\pm(t), \quad \tilde{W}_\pm(t) = rt + \tilde{d}_\pm(t),$$

where $d_\pm$ and $\tilde{d}_\pm$ are 1-periodic functions. Their inverses are $W_\pm^{-1}(s) = r^{-1}s + b_\pm(s)$ and $\tilde{W}_\pm^{-1}(s) = r^{-1}s + \tilde{b}_\pm(s)$ with r-periodic functions $b_\pm$ and $\tilde{b}_\pm$. Now relations (2.4.2) take on the form

$$W_\pm(H(t)) = \tilde{W}_\pm(t) + \mu\,,$$

where $H(t)$ is a diffeomorphism commuting with the standard shift. Therefore

$$H(t) = W_+^{-1}(\tilde{W}_+(t) + \mu) = W_-^{-1}(\tilde{W}_-(t) + \mu), \tag{2.4.4}$$

and we find that

$$v_F(t) = W_-W_+^{-1}(t) - t, \quad v_{\tilde{F}}(t) = \tilde{W}_-\tilde{W}_+^{-1}(t) - t$$

are r-periodic C^k-functions satisfying

$$v'_F(t) + 1 \neq 0, \quad v'_{\tilde{F}}(t) + 1 \neq 0,$$

while (2.4.4) states

$$v_{\tilde{F}}(t) = v_F(t + \mu). \tag{2.4.5}$$

As a result we obtain the following development of Theorem 2.12 for $G = G_{r,0}$.

Corollary 2.10. *i) If diffeomorphisms $F, \tilde{F} \in C(k, G_{r,0})$ are C^1-conjugate on $\mathbb{T}$, then (2.4.5) holds with some $\mu \in \mathbb{R}$.*

ii) If (2.4.5) holds for diffeomorphisms $F, \tilde{F} \in C(k, G_{r,0})$, then they are C^k-conjugate on $\mathbb{T}$.

iii) For every r-periodic C^k-function v satisfying $v'(t) + 1 \neq 0$, $t \in \mathbb{R}$, there exists a diffeomorphism $F \in C(k, G_{r,0})$ such that $v_F = v$.

Proof. i) has already been shown. To prove ii) we note that (2.4.5) is equivalent to (2.4.4) which in turn defines a C^k-diffeomorphism $H(t)$ commuting with the standard shift. Therefore (2.4.2) is fulfilled and by Theorem 2.12 we conclude that F and $\tilde{F}$ are C^k-conjugate.

To prove iii) we note that C^k-functions $d_+(t) = v(rt)$, $d_-(t) = v(rt + v(rt))$ are 1-periodic and $W_+(t) = rt + v(rt)$, $W_-(t) = W_+(t) + v(W_+(t))$ are C^k-diffeomorphisms of the real line. If we set $D_\pm(t) = (W_\pm(t))^{-1/r}$, then $G_{r,0}D_\pm(t) = (W_\pm(t) + r)^{-1/r}$ and since $W_+(t) + r = r(t+1) + v(rt) = W_+(t+1)$, $W_-(t) + r = W_+(t) + r + v(W_+(t) + r) = W_-(t+1)$, we have $G_{r,0}D_\pm(t) = D_\pm(t+1)$. Therefore the functions $D_\pm$ are representatives of a class $c \in J(k, G_{r,0})$, and according to Theorem 2.12 they correspond to some C^k-diffeomorphism $F \in C(k, G_{r,0})$. □

It was already proved in Theorem 2.10 that there are no functional invariants (similar to functions v_F) which are obstacles to the topological classification of diffeomorphisms from class $C(k, G_{r,0})$. All such diffeomorphisms are not only globally topologically conjugate, but they are locally C^k-conjugate at every point $z \in \mathbb{T}$. Nevertheless, as stated by Corollary 2.10 they may not be C^1-conjugate on $\mathbb{T}$, since condition (2.4.5) may fail. Moreover, the same Corollary 2.10 implies that $F \in C(k, G_{r,0})$ is C^1-conjugate with some C^{k+1}-diffeomorphism if and only if $v_F \in C^{k+1}$. It means that it is impossible to increase the smoothness class of C^k-diffeomorphism F by a C^1-transformation $F \to \Phi F \Phi^{-1}$ if $v_F \notin C^{k+1}$.

2.4.3 Smooth classification of diffeomorphisms with several hyperbolic fixed points

We have already shown in the previous section that there exists an essential difference between smooth classification of diffeomorphisms with a unique fixed point on the real line and on the circle. Namely, in the case of $M = \mathbb{R}$ the class of smooth equivalence of a diffeomorphism is completely determined by its local structure while for $M = \mathbb{T}$ additional global invariants arise. We will see now that in the presence of several fixed points similar global invariants appear on the real line as well. Construction of such invariants may be carried out in the same way in both cases, and here we restrict ourselves to the case $M = \mathbb{R}$ only.

Let $F : \mathbb{R} \to \mathbb{R}$ be a C^k-diffeomorphism, $k \geq 1$, with hyperbolic fixed points only, and let $\mathrm{Fix}F = \{x_1, x_2, \ldots, x_n\}$, $n \geq 2$. Then F preserves the orientation on $\mathbb{R}$ and repelling and attracting fixed points alternate. To be definite, let

$$\lambda_{2j-1} = F'(x_{2j-1}) < 1 < \lambda_{2j} = F'(x_{2j}), \quad j = 1, 2, \cdots .$$

The intervals

$$U_1 = (-\infty, x_2); \quad U_j = (x_{j-1}, x_{j+1}), \quad 2 \le j \le n-1; \quad U_n = (x_{n-1}, +\infty),$$

are invariant with respect to F, and for each j, $1 \le j \le n$, the restriction $F|_{U_j} : U_j \to U_j$ is a C^k-diffeomorphism with the unique fixed point x_j.

Let us denote by $C(k; \lambda_1, \ldots, \lambda_n)$ the set of all C^k-diffeomorphisms F with derivatives $F'(x_j) = \lambda_j$ at fixed points x_j, $1 \le j \le n$. We assume that for $k = 1$ the mapping $F|_{U_j}$ is C^1-conjugate with its linear part $\Lambda_j(x) = \lambda_j x$; the C^k-conjugacy for $k \ge 2$ follows from Theorem 2.5.

Let $\Phi_j : U_j \to \mathbb{R}$, $\Phi_j(x_j) = 0$, $1 \le j \le n$, be an orientation-preserving C^k-diffeomorphism conjugating $F|_{U_j}$ with Λ_j. Loosely speaking we can say that F is glued together from linear mappings using *gluing* C^k-diffeomorphisms

$$G_j = \Phi_{j+1}\Phi_j^{-1} : \Phi_j(U_j \bigcap U_{j+1}) \to \Phi_{j+1}(U_j \bigcap U_{j+1}), \quad 1 \le j \le n-1.$$

Since $U_j \bigcap U_{j+1} = (x_j, x_{j+1})$, we have

$$\Phi_j(U_j \bigcap U_{j+1}) = \mathbb{R}_+, \quad \Phi_{j+1}(U_j \bigcap U_{j+1}) = \mathbb{R}_-.$$

Hence

$$G_j : \mathbb{R}_+ \to \mathbb{R}_-$$

is an orientation-preserving diffeomorphism. The identity

$$F(x) = \Phi_j^{-1}\Lambda_j\Phi_j(x) = \Phi_{j+1}^{-1}\Lambda_{j+1}\Phi_{j+1}(x), \quad x \in U_j \bigcap U_{j+1},$$

implies

$$G_j(\lambda_j x) = \lambda_{j+1} G_j(x), \quad x \in \mathbb{R}_+. \tag{2.4.6}$$

Therefore

$$G_j(x) = -x^{\alpha_j}\tau_j\left(\frac{\ln x}{\ln \lambda_j}\right), \quad \alpha_j = \frac{\ln \lambda_{j+1}}{\ln \lambda_j}, \tag{2.4.7}$$

where $\tau_j(t) > 0$ is a 1-periodic C^k-function. Since $G_j'(x) > 0$, $x \in \mathbb{R}_+$, we have

$$\alpha_j \tau_j(t) + \frac{1}{\ln \lambda_j}\tau'(t) < 0.$$

The above construction relates to every $F \in C(k; \lambda_1 \ldots \lambda_n)$ a collection $\{G_j\}_1^{n-1}$ of C^k-diffeomorphisms $G_j : \mathbb{R}_+ \to \mathbb{R}_-$ satisfying (2.4.6). For a fixed F this collection depends on the choice of conjugating diffeomorphisms Φ_j. If $\tilde{\Phi}_j$, $1 \le j \le n$, is another set of such reducing C^k-diffeomorphisms, then every diffeomorphism $\tilde{\Phi}_j\Phi_j^{-1} : \mathbb{R} \to \mathbb{R}$, $1 \le j \le n$, commutes with the linear transform Λ_j and is a linear function itself, i.e., $\tilde{\Phi}_j\Phi_j^{-1}(t) = a_j t$, $a_j > 0$. Therefore the gluing collections $\{G_j\}_1^{n-1}$ and $\{\tilde{G}_j\}_1^{n-1}$ are connected by the relations

$$\tilde{G}_j(a_j x) = a_{j+1}G_j(x), \quad 1 \le j \le n-1. \tag{2.4.8}$$

Two collections $\{G_j\}_1^{n-1}$ and $\{\tilde{G}_j\}_1^{n-1}$ of C^k-diffeomorphisms from $\mathbb{R}_+$ onto $\mathbb{R}_-$ satisfying (2.4.6) are said to be *equivalent* if (2.4.8) holds with some $a_j > 0$, $1 \le j \le n-1$. We denote by $J(k;\lambda_1, \dots, \lambda_n)$ the set of equivalence classes of such collections. The above construction determines a mapping

$$I : C(k;\lambda_1, \dots, \lambda_n) \to J(k;\lambda_1, \dots, \lambda_n).$$

The following proposition is similar to Corollary 2.10.

Theorem 2.13. *i) If diffeomorphisms $F, \tilde{F} \in C(k;\lambda_1, \dots, \lambda_n)$ are C^1-conjugate, then $I(F) = I(\tilde{F})$.*

ii) If $I(F) = I(\tilde{F})$, then F and $\tilde{F}$ are C^k-conjugate.

iii) The mapping I is surjective: for every $c \in J(k;\lambda_1, \dots, \lambda_n)$ there exists a diffeomorphism $F \in C(k;\lambda_1, \dots, \lambda_n)$ such that $I(F) = c$.

Proof. Assume $\tilde{F} = \Psi F \Psi^{-1}$ where $\Psi \in C^1$. If U_j, $1 \le j \le n$, are the same invariant intervals for F as before, then $\tilde{U}_j = \Psi(U_j)$, $1 \le j \le n$, are invariant intervals for $\tilde{F}$, and

$$\tilde{\Phi}_j^{-1}\Lambda_j\tilde{\Phi}_j\Psi|_{U_j} = \Psi\Phi_j^{-1}\Lambda_j\Phi_j, \quad 1 \le j \le n,$$

where Φ_j and $\tilde{\Phi}_j$ are reducing diffeomorphisms for F and $\tilde{F}$, respectively. It means that every diffeomorphism $H_j = \tilde{\Phi}_j\Psi\Phi_j^{-1} : \mathbb{R} \to \mathbb{R}$ commutes with the linear mapping Λ_j and hence it is linear itself: $H_j(x) = a_j x$, $a_j > 0$. Therefore (2.4.8) is fulfilled, which means that $I(F) = I(\tilde{F})$, proving *i*).

To prove *ii*) let F and $\tilde{F}$ be diffeomorphisms such that $I(F) = I(\tilde{F})$. Then (2.4.8) holds and the relations

$$\Psi|_{U_j} = \tilde{\Phi}_j^{-1}H_j\Phi_j, \quad H_j(x) = a_j x, \quad 1 \le j \le n,$$

determine a well-defined C^k-diffeomorphism $\Psi : \mathbb{R} \to \mathbb{R}$ conjugating F and $\tilde{F}$.

Let now a collection $\Omega = \{G_j\}_1^{n-1}$ of C^k-diffeomorphisms $G_j : \mathbb{R}_+ \to \mathbb{R}_-$ be given satisfying (2.4.6). To find a diffeomorphism F from statement *iii*) we will find C^k-diffeomorphisms Φ_j such that $G_j = \Phi_{j+1}\Phi_j^{-1}$, $j = 1, \dots, n-1$.

First, we fix points $y_1 < y_2 < \dots < y_n \in \mathbb{R}$, introduce the corresponding intervals $V_j = (y_{j-1}, y_{j+1})$, $2 \le j \le n-1$, $V_1 = (-\infty, y_2)$, $V_n = (y_{n-1}, +\infty)$, and choose a homeomorphism $\Psi_1 : V_1 \to \mathbb{R}$ such that $\Psi_1(-\infty, y_1) = \mathbb{R}_-$, $\Psi_1(y_1, y_2) = \mathbb{R}_+$. The function $G_1(x)$ is representable by (2.4.7) with $\alpha_1 < 0$, $\tau_1(t) > 0$ and the restriction of $G_1\Psi_1$ to the interval (y_1, y_2) maps it onto $\mathbb{R}_-$ with $G_1\Psi_1(t) \to 0$ as $t \to y_2$. Hence we can extend $G_1\Psi_1$ as a homeomorphism $\Psi_2 : V_2 \to \mathbb{R}$.

At the second step we use the homeomorphism $G_2\Psi_2 : (y_2, y_3) \to \mathbb{R}_-$ to define a homeomorphism $\Psi_3 : V_3 \to \mathbb{R}$ coinciding with $G_2\Psi_2$ on the interval (y_2, y_3). Repeating this construction step by step we obtain a collection of homeomorphisms $\{\Psi_j\}_1^n$, $\Psi_j : V_j \to \mathbb{R}$, $\Psi_j(y_j) = 0$, such that

$$G_j = \Psi_{j+1}\Psi_j^{-1},$$

but not necessarily belonging to the given class of smoothness. To "correct" this collection note that the collection $\{\Psi_j\}_1^n$ defines an atlas of a C^k-manifold Z homeomorphic to $\mathbb{R}$. Because of the uniqueness of the smooth structure on $\mathbb{R}$ there exists a C^k-diffeomorphism $H : \mathbb{R} \to Z$, which means that the mappings

$$\Phi_j = \Psi_j H : H^{-1}(V_j) \to \mathbb{R}, \quad 1 \le j \le n,$$

are C^k-diffeomorphisms. Now we have

$$\Phi_{j+1}\Phi_j^{-1} = \Psi_{j+1}\Psi_j^{-1} = G_j, \quad 1 \le j \le n,$$

and if we set $U_j = H^{-1}(V_j)$, $x_j = H^{-1}(y_j)$, $1 \le j \le n$, then the relations

$$F|_{U_j} = \Phi_j^{-1}\Lambda_j\Phi_j, \quad 1 \le j \le n,$$

define a C^k-diffeomorphism with the fixed points $x_1, \dots x_n$ such that $I(F)$ is generated by the collection $\{G_j\}_1^{n-1}$. □

The statement *iii)* implies that the set $C(k; \lambda_1, \dots, \lambda_n)$ contains smoothly non-conjugate diffeomorphisms.

Corollary 2.11. *There exist analytic diffeomorphisms* $F, \tilde{F} \in C(\aleph; \lambda_1, \dots, \lambda_n)$ *which are not* C^1*-conjugate.*

Proof. Indeed, let us consider the "trivial" collection $\Omega = \{-x^{\alpha_j}\}_1^{n-1}$ corresponding to the functions $\tau_j(t) \equiv 1$ from (2.4.7) and an analytic diffeomorphism F generated by this collection. If $\tilde{\Omega} = \{\tilde{G}_j\}_1^{n-1}$ is another collection with at least one function $\tau_i(t)$ not constant, e.g.,

$$\tau_j(t) = 1 + \varepsilon \sin 2\pi t,$$

with sufficiently small $|\varepsilon| \neq 0$, then $\tilde{\Omega}$ and Ω are not equivalent, and hence any analytic diffeomorphism $\tilde{F}$ generated by $\tilde{\Omega}$ is not C^1-conjugate with F. □

Similar to the situation with the unique fixed point on $\mathbb{T}$ discussed at the end of the previous section, we can state the following criterion.

Corollary 2.12. *A diffeomorphism* $F \in C(k; \lambda_1, \dots, \lambda_n)$ *is* C^1*-conjugate with a* C^p*-diffeomorphism if and only if* $I(F) \in J(p; \lambda_1, \dots, \lambda_n)$,

In other words, the corresponding gluing diffeomorphisms G_j must belong to the class C^p.

2.4.4 Another approach to smooth classification

The method of smooth classification described in Sections 2.3.2 and 2.3.3 may be extended to the case of several and even an infinite number of isolated fixed points. Similar to a local situation this approach uses the straightening of a given diffeomorphism on the complement to the set of fixed points without fixing local normal forms.

Let, as before, $F : \mathbb{R} \to \mathbb{R}$ be a C^k-diffeomorphism, $\operatorname{Fix} F = \{x_1, \dots, x_n\}$, $n \geq 2$, and let

$$V_0 = (-\infty, x_1), \quad V_j = (x_j, x_{j+1}), \quad 1 \leq j \leq n-1, \quad V_n = (x_n, +\infty).$$

Since F is a C^k-diffeomorphism of V_j, $0 \leq j \leq n$, onto itself without fixed points, by virtue of Corollary 2.2 every restriction $F|_{V_j}$, $0 \leq j \leq n$, is C^k-conjugate with the standard shift. If $\Psi_j : V_j \to \mathbb{R}$, $0 \leq j \leq n$, are conjugating diffeomorphisms, then

$$F(x) = \Psi_j^{-1}(\Psi_j(x) + 1), \quad x \in V_j, \ 0 \leq j \leq n. \tag{2.4.9}$$

As in Section 2.3 we call $\{\Psi_j\}_0^n$ the straightening diffeomorphisms for F.

Theorem 2.14. *Let F and $\tilde{F}$ be two diffeomorphisms with collections $\{\Psi_j\}_0^n$ and $\{\tilde{\Psi}_j\}_0^n$ of straightening diffeomorphisms, respectively. Then F and $\tilde{F}$ are C^p-conjugate if and only if there exist the C^p-diffeomorphism $\Phi : \mathbb{R} \to \mathbb{R}$ and C^p-diffeomorphisms $H_j : \mathbb{R} \to \mathbb{R}$, $0 \leq j \leq n$, commuting with the standard shift such that*

$$\tilde{\Psi}_j \Phi = H_j \Psi_j, \quad 0 \leq j \leq n.$$

The latter conditions mean that the piecewise mapping Φ defined by

$$\Phi(x) = \tilde{\Psi}_j^{-1} H_j \Psi_j(x), \quad x \in V_j, \ 0 \leq j \leq n,$$

is extendable as a global C^p-diffeomorphism of the real line.

We omit the proof of Theorem 2.14, since it repeats, besides evident technical details, the arguments used to prove Theorem 2.8.

Let us apply the described approach to C^1-diffeomorphisms of the line with hyperbolic fixed points.

Let, as above,

$$0 < \lambda_{2j-1} = F'(x_{2j-1}) < 1 < \lambda_{2j} = F'(x_{2j}) \quad j = 1, 2, \dots$$

be derivatives of F at the fixed points $x_1 < x_2 < \cdots$. Then for the straightening diffeomorphisms $\Psi_j : V_j \to \mathbb{R}$ we have $\Psi'_{2j}(x) > 0$ and $\Psi'_{2j+1}(x) < 0$, $j = 0, 1, 2, \dots$. Using (2.4.9) we find that $\theta_0(t) \equiv (\Psi_0^{-1}(t))'$ is a positive continuous function satisfying conditions

$$\lim_{t \to +\infty} \frac{\theta_0(t+1)}{\theta_0(t)} = \lambda_1, \quad \int_{-\infty}^{+\infty} \theta_0(s) ds = +\infty. \tag{2.4.10}$$

It is evident that the representation

$$\Psi_0^{-1}(t) = x_1 - \int_t^{+\infty} \theta_0(s)ds \tag{2.4.11}$$

holds. Furthermore,

$$\Psi_k^{-1}(t) = x_k + \begin{cases} \Delta x_k \int_t^{+\infty} \theta_k(s)ds, & k = 2j-1, \quad 1 \le 2j-1 < n, \\ \Delta x_k \int_{-\infty}^{t} \theta_k(s)ds, & k = 2j, \qquad 2 \le 2j < n, \end{cases} \tag{2.4.12}$$

where $\Delta x_k = x_{k+1} - x_k$ and continuous positive functions $\theta_k(s)$ satisfy conditions

$$\lim_{t\to+\infty} \frac{\theta_k(t+1)}{\theta_k(t)} = \lambda_k, \quad \lim_{t\to-\infty} \frac{\theta_k(t+1)}{\theta_k(t)} = \lambda_{k+1},$$

$$\int_{-\infty}^{\infty} \theta_k(s)ds = 1, \quad 1 \le k \le n-1. \tag{2.4.13}$$

In addition,

$$\Psi_n^{-1}(t) = \begin{cases} x_n + \int_t^{+\infty} \theta_n(s)ds, & n = 2q+1, \\ x_n + \int_{-\infty}^{t} \theta_n(s)ds, & n = 2q, \end{cases} \tag{2.4.14}$$

where $\theta_n(t) > 0$ is a continuous function and

$$\lim_{t\to(-1)^{n+1}\infty} \frac{\theta_n(t+1)}{\theta_n(t)} = \lambda_n, \quad \int_{-\infty}^{\infty} \theta_n(s)ds = \infty. \tag{2.4.15}$$

The above collection $\{\theta_j\}_0^n$ of continuous positive functions satisfies (2.4.10), (2.4.13) and (2.4.15) and may be regarded as a system of functional coordinates for F.

On the other hand, given such a collection and a system $x_1 < x_2 < \cdots < x_n$ of points on the real line, we can define the system of C^1-diffeomorphisms $\{\Psi_j\}_0^\infty$ using (2.4.11), (2.4.12) and (2.4.14) and check that (2.4.9) defines a C^1-diffeomorphism $F : \mathbb{R} \to \mathbb{R}$ with $\operatorname{Fix} F = \{x_j\}_1^n$, $F'(x_j) = \lambda_j$, $1 \le j \le n$, and a collection of functional coordinates $\{\theta_j\}_0^n$.

Let $\tilde{F}$ be another C^1-diffeomorphism with the same fixed points $\{x_j\}_1^n$ and with the same derivatives $\lambda_j = \tilde{F}'(x_j)$, $1 \leq j \leq n$. Applying Theorem 2.14 we obtain the following criterion of C^1-conjugacy in terms of functional coordinates $\{\theta_j\}_0^n$ and $\{\tilde{\theta}_j\}_0^n$.

Corollary 2.13. *The diffeomorphisms F and $\tilde{F}$ are C^1-conjugate if and only if there exist C^1-diffeomorphisms $H_j : \mathbb{R} \to \mathbb{R}$ commuting with the standard shift such that*

$$\lim_{t\to+\infty} \frac{\tilde{\theta}_{2j}(H_{2j}(t))H'_{2j}(t)}{\theta_{2j}(t)} = \lim_{t\to+\infty} \frac{\tilde{\theta}_{2j+1}(H_{2j+1}(t))H'_{2j+1}(t)}{\theta_{2j+1}(t)}, \quad 2j+1 \leq n,$$

$$\lim_{t\to-\infty} \frac{\tilde{\theta}_{2j-1}(H_{2j-1}(t))H'_{2j-1}(t)}{\theta_{2j-1}(t)} = \lim_{t\to-\infty} \frac{\tilde{\theta}_{2j}(H_{2j}(t)H'_{2j}(t)}{\theta_{2j}(t)}, \quad 2j-1 \geq 1,$$

where all limits are finite and different from zero.

The following example illustrates the above statements.

Example 2.9. Let $0 < \lambda < 1 < \mu$ and

$$\theta_0(t) = e^{t\ln\lambda}, \quad \theta_1(t) = ce^{t\alpha(t)}, \quad \theta_2(t) = e^{t\ln\mu}$$

where

$$\alpha(t) = \frac{1}{2}\ln\lambda\mu - \frac{1}{\pi}\ln\frac{\mu}{\lambda}\arctan t, \quad c^{-1} = \int_{-\infty}^{\infty} e^{s\alpha(s)}ds.$$

Diffeomorphisms $\Psi_j(x)$, $j = 0,1,2$, defined by their inverse

$$\Psi_0^{-1}(t) = \frac{e^{t\ln\lambda}}{\ln\lambda}, \quad \Psi_1^{-1}(t) = \int_t^{\infty} \theta_1(s)ds, \quad \Psi_2^{-1}(t) = 1 + \frac{e^{t\ln\mu}}{\ln\mu},$$

map $V_0 = (-\infty,0)$, $V_1 = (0,1)$, $V_2 = (1,+\infty)$, respectively, onto $(-\infty,\infty)$, and relations (2.4.9) with $n = 2$ define a C^1-diffeomorphism with fixed points $x_1 = 0$, $x_2 = 1$, derivatives $F'(0) = \lambda$; $F'(1) = \mu$, and functional coordinates $\theta_j(t)$, $j = 0,1,2$.

Let $\tilde{F}$ be a C^1-diffeomorphism with the same fixed points, the same derivatives $\tilde{F}'(0) = \lambda$, $\tilde{F}'(1) = \mu$, and with functional coordinates $\tilde{\theta}_j(t)$, $j = 0,1,2$. Now Corollary 2.13 implies (cf., Corollary 2.6) that F and $\tilde{F}$ are C^1-conjugate if and only if there exist continuous 1-periodic functions $p_0(t) > 0$, $p_2(t) > 0$ and a smooth 1-periodic function $\tau(t)$, $\tau'(t) \neq -1$ such that

1. $\tilde{\theta}_0(t) = \theta_0(t)p_0(t)(1 + o(1))$, $t \to +\infty$.
2. $\tilde{\theta}_2(t) = \theta_2(t)p_2(t)(1 + o(1))$, $t \to -\infty$.

3. The limits

$$\lim_{t\to+\infty} \frac{\theta_1(t)e^{\tau(t)\ln\lambda}(1 + \tau'(t))}{\tilde{\theta}_1(t)},$$

$$\lim_{t\to-\infty} \frac{\theta_1(t)e^{\tau(t)\ln\mu}(1 + \tau'(t))}{\tilde{\theta}_1(t)}$$

exist and are different from zero.

2.5 One-dimensional flows and vector fields

Let us recall that a *flow* on M is a one-parametric group of homeo- or diffeomorphisms $F^t : M \to M$, $t \in \mathbb{R}$. This means that

$$F^{t+s}(x) = F^t(F^s(x)), \quad s,t \in \mathbb{R},\ x \in M.$$

In particular, $F^0(x) = x$, $(F^t(x))^{-1} = F^{-t}(x)$.

Every diffeomorphism $F : M \to M$ defines a group of diffeomorphisms $F^n : M \to M$, $n \in \mathbb{Z}$. Such a "discrete" flow is usually called a *cascade*. All homeomorphisms of a flow (cascade) commute: $F^t(F^s) = F^{s+t} = F^s(F^t)$.

The simplest example of a flow on the real line is given by a *shift* $F^t(x) = x + \alpha(t)$ where $\alpha : \mathbb{R} \to \mathbb{R}$ is a function satisfying the *Cauchy equation*

$$\alpha(t + s) = \alpha(t) + \alpha(s).$$

It is well known that under minimal assumptions (local boundedness, measurability) a solution of the Cauchy equation is a linear function $\alpha(t) = ct$. On the other hand, there are many other solutions which may be constructed using any Hamel basis of the linear space $\mathbb{R}$ over the field of rational numbers.

In particular, the shift $F_0^t(x) = x + t$ is called the *standard flow*.

We consider continuous, smooth and analytic flows. This means that the mapping $(x,t) \to (t, F^t(x))$ of the plane is a diffeomorphism from the corresponding smoothness class.

Two flows F^t and $\tilde{F}^t$ are called C^k*-conjugate* if

$$\Phi(F^t) = \tilde{F}^t(\Phi), \quad t \in \mathbb{R},$$

where $\Phi : M \to M$ is a C^k-diffeomorphism. Let us note that cascades $\{F^n\}$ and $\{\tilde{F}^n\}$ are conjugate as soon as their *generators* F and $\tilde{F}$ are conjugate. The latter is not true for flows: generally speaking, the conjugacy of homeomorphisms $F^t|_{t=1}$ and $\tilde{F}^t|_{t=1}$ does not imply the conjugacy of the corresponding flows.

Homeomorphisms F^t composing a flow possess some specific properties. For instance, they preserve the orientation, which follows from the continuity of $F^t(x)$ with respect to t and the equality $F^0(x) = x$. In particular, it means that if a homeomorphism F does not preserve the orientation, it may not be embedded in a flow: there is no flow F^t such that $F = F^t|_{t=t_0}$. Another consequence of preserving the orientation on $\mathbb{R}$ is that for every $t \in \mathbb{R}$ all non-wandering points of F^t are fixed.

Theorem 2.15. *Given a C^k-flow $F^t : \mathbb{R} \to \mathbb{R}$, assume that there exists at least one diffeomorphism F^{t_0} without fixed points. Then F^t is C^k-conjugate with the standard flow.*

Proof. According to Corollary 2.2, the diffeomorphism F^{t_0} is C^k-conjugate with the standard shift and hence we can assume $F^{t_0}(x) = x + t_0$, $t_0 \neq 0$. Since F^{t_0} commutes with F^t, we have

$$F^t(x + t_0) = F^t(x) + t_0.$$

The function

$$h(t, x) = F^{-t}(x) + t$$

is a C^k-diffeomorphism of $\mathbb{R}$ for each t, and

$$h(t + t_0, x) = h(t, x).$$

The mapping

$$\Phi(x) = \frac{1}{t_0} \int_0^{t_0} h(s, x)ds$$

is a C^k-diffeomorphism of the real line as well. Let us show that it conjugates F^t with the standard flow. Indeed,

$$\Phi(x) + t = \frac{1}{t_0} \int_0^{t_0} (h(s, x) + t)ds = \frac{1}{t_0} \int_0^{t_0} h(s + t, F^t(x))ds.$$

Since $h(t, x)$ is t_0-periodic with respect to t, the substitution $s + t = u$ yields

$$\Phi(x) + t = \frac{1}{t_0} \int_t^{t+t_0} h(u, F^t x)du = \frac{1}{t_0} \int_0^{t_0} h(u, F^t(x))du.$$

Hence

$$\Phi(x) + t = \Phi(F^t(x)).$$

□

Every C^k-flow F^t, $k \geq 1$, generates a *vector field* $v = v(x)\partial/\partial x$, which is a linear operator acting in the space of smooth functions $\varphi(x)$ according to the formula

$$(v\varphi)(x) = v(x)\frac{\partial \varphi}{\partial x}, \quad v(x) = \frac{dF^t(x)}{dt}|_{t=0}.$$

It is easy to see that the function $x(t) = F^{t-t_0}(x_0)$ is a solution of the Cauchy problem

$$\dot{x} = v(x), \quad x(t_0) = x_0. \tag{2.5.1}$$

Conversely, let $v(x)$ be a smooth function. According to the Existence and Uniqueness Theorem, for any point (t_0, x_0) there exists the unique solution $x(t)$ of (2.5.1) defined in a neighborhood of t_0. Denote by $F^t(x_0)$ the solution $x(t)$ with the initial condition $x(0) = x_0$. If t, s and $t + s$ belong to the domain of existence and uniqueness of the solution, then $F^t F^s(x) = F^{t+s}(x)$. In this sense any vector field v or differential equation (2.5.1) generate a *local flow* $F^t(x)$ defined for every $x \in M$ on some interval $U(x) \in M$. The compactness of $M = \mathbb{T}$ implies that, in fact, the solution is defined everywhere and $U(x) = \mathbb{T}$. If $M = \mathbb{R}$, then $U(x)$ may be a finite interval. For example, for $v(x) = x^2 + 1$ we have $F^t(x) = \tan(t + \arctan x)$ and hence

$$U(x) = (-\frac{\pi}{2} - \arctan x, \frac{\pi}{2} - \arctan x).$$

If $v(x) = x^{k+1}$, $k \geq 1$, then $F^t(x) = x(1 - ktx^k)^{-1/k}$, and

$$U(x) = (-\infty, \frac{1}{kx^k}), \quad x \neq 0; \quad U(0) = \mathbb{R},$$

if k is even, and

$$U(x) = (-\infty, \frac{1}{kx^k}), \quad x > 0; \quad U(0) = \mathbb{R}; \quad U(x) = (\frac{1}{kx^k}, +\infty), \quad x < 0,$$

if k is odd.

A vector field on $\mathbb{R}$ is said to be *complete* if the corresponding flow F^t is defined for all $t \in \mathbb{R}$, which means $U(x) = \mathbb{R}$, $x \in \mathbb{R}$. The relation

$$v(x) = \frac{dF^t(x)}{dt}|_{t=0}, \quad x \in M,$$

defines a bijection between flows and complete vector fields. The following are some examples of complete vector fields on $\mathbb{R}$:

1. $v(x) = c$, $F^t(x) = x + ct$;
2. $v(x) = \alpha x$, $F^t(x) = e^{\alpha t}x$.

A smooth transformation $x \to \Phi(x)$ reduces equation (2.5.1) to a similar equation with the function

$$\tilde{v}(x) = (\Phi'(x))^{-1} v(\Phi(x)), \tag{2.5.2}$$

and Φ conjugates the corresponding flows $\tilde{F}^t$ and F^t.

Relation (2.5.2) between vector fields v and $\tilde{v}$ will be also called the *conjugacy*. Of course, the conjugacy, if it exists, presumes that Φ is a smooth transformation. On the other hand, the relation $\tilde{F}^t = \Phi^{-1} F^t \Phi$ requires the mapping Φ to be just a continuous homeomorphism. In this sense the latter relation defines the *topological conjugacy* of vector fields.

A point x_0 is called *non-singular* for a vector field v if $v(x_0) \neq 0$. Every vector field may be straightened in a neighborhood of such a point: there exists a local invertible transformation $x \to \Phi(x)$ reducing v to the constant vector field $\tilde{v}(x) = 1$.

More exactly, let us assume $v(x) \neq 0, x \in (a,b)$, choose $\xi \in (a,b)$ and set

$$\Phi(x) = \int_{\xi}^{x} \frac{dt}{v(t)}.$$

This function is a C^k-diffeomorphism of (a,b) onto the interval with the end-points

$$a_1 = \int_{\xi}^{a} \frac{dt}{v(t)}, \quad b_1 = \int_{\xi}^{b} \frac{dt}{v(t)}.$$

Since $\Phi'(x)v(x) = 1$, the diffeomorphism Φ conjugates v with the constant field. In particular, if all points $x \in \mathbb{R}$ are non-singular for v, then the function $\Phi(x)$ is determined for all $x \in \mathbb{R}$. It is a C^k-diffeomorphism of $\mathbb{R}$ onto $\mathbb{R}$ if and only if both integrals

$$\int_{-\infty} \frac{dt}{v(t)}, \quad \int^{+\infty} \frac{dt}{v(t)}$$

diverge. Therefore the divergence of the above integrals is a necessary and sufficient condition for a non-singular vector field on the real line to be conjugate with a constant field. If the integral

$$\chi(v) = \int_{-\infty}^{\infty} \frac{dt}{v(t)}$$

converges, then the number $\chi(v)$ is an invariant with respect to the global conjugacy.

2.5.1 Classification of vector fields in a neighborhood of a singular point

A point $x_0 \in M$ is called *singular* for a vector field v if $v(x_0) = 0$. Singular points of a vector field v correspond to fixed points of its flow: $F^t(x_0) = x_0$, $t \in \mathbb{R}$.

The dynamical behavior of a flow in a neighborhood of an isolated singular point is the same as that of a cascade in a neighborhood of its fixed point. Similar to Section 2.3 we can introduce the notions of attracting, repelling and overflow points and prove, as in Lemma 2.1, that there exists a pair of local straightening diffeomorphisms.

A singular point x_0 is called *hyperbolic* if $v'(x_0) \neq 0$. Since

$$\frac{dF^t(x)}{dx}|_{x=x_0} = e^{tv'(x_0)},$$

a hyperbolic point of a vector field v corresponds to a hyperbolic fixed point of diffeomorphisms F^t.

The following statement is an analogue of Theorem 2.5 for vector fields.

Theorem 2.16. *A local C^k-vector field v, $k \geq 2$, in a neighborhood of a hyperbolic singular point $x_0 \in \mathbb{R}$ is C^k-conjugate with the linear field $\tilde{v}(x) = \alpha(x - x_0)\partial/\partial x$, $\alpha = v'(x_0)$.*

Proof. Let F^t be a local flow generated by v in a neighborhood V of x_0. As above, denote by $U(x)$ the domain of existence of $F^t(x)$. Without loss of generality we may assume that there exists $t_0 > 0$ such that $2t_0 \in U(x)$ for all x from V. According to Theorem 2.5 we can suppose that

$$F^{t_0}(x) = e^{\alpha t_0}(x - x_0) + x_0, \quad x \in V.$$

Set $G^s(x) = e^{\alpha s}(x - x_0) + x_0$ and

$$h(s,x) = G^s F^{-s}(x), \quad s \in (-t_0, t_0), \quad x \in V.$$

Since $s + t_0 \in U(x)$, we have

$$h(s + t_0, x) = e^{\alpha(s+t_0)}(F^{-t_0}F^{-s}(x) - x_0) + x_0$$

$$= e^{\alpha(s+t_0)}(e^{-\alpha t_0}(F^{-s}(x) - x_0)) + x_0 = h(s,x),$$

and

$$\frac{\partial h(s,x)}{\partial x}|_{x=x_0} = e^{\alpha s}\frac{\partial F^{-s}(x)}{\partial x}|_{x=x_0} = e^{\alpha s - v'(x_0)s} = 1.$$

We set

$$\Phi(x) = \frac{1}{t_0}\int_0^{t_0} h(s,x)ds$$

and obtain a local C^k-diffeomorphism in V, $\Phi(x_0) = x_0$. Since $h(s,x)$ is t_0-periodic,

$$G^t\Phi(x) = \frac{1}{t_0}\int_0^{t_0} G^t h(s,x)ds$$

$$= \frac{1}{t_0}\int_0^{t_0} h(s+t, F^t(x))ds = \frac{1}{t_0}\int_0^{t_0} h(u, F^t(x))du.$$

Therefore

$$G^t\Phi(x) = \Phi F^t(x),$$

which means that Φ conjugates G^t and F^t, and (2.5.2) follows. □

Similar to Theorem 2.5, the condition $k \geq 2$ is essential. Indeed, it is easy to check (cf., Example 2.5) that the C^1-vector field

$$v(x) = \alpha x + \int_0^x \frac{dt}{\ln|t|}, \quad |x| < \frac{1}{2}, \ \alpha < 0,$$

is not smoothly conjugate with the linear vector field $\tilde{v}(x) = \alpha x$. We will construct in Section 2.5.2 functional coordinates of C^1-flows which are similar to coordinates of local diffeomorphisms introduced in Section 2.3.2.

Let us now pass to a local classification of vector fields at non-hyperbolic singular points. We start with a problem of the formal equivalence.

A *formal vector field* is an expression

$$v = v(x)\frac{\partial}{\partial x}$$

which defines a differential operator

$$(v\varphi)(x) = v(x)\varphi'(x), \quad \varphi \in \mathbb{R}[[x]].$$

Formal vector fields v and $\tilde{v}$ are said to be *conjugate* if

$$\tilde{v}(x) = (\Phi'(x))^{-1}v(\Phi(x))$$

with some invertible formal mapping $\Phi(x)$.

Repeating arguments used to prove Theorem 2.6, we obtain normal forms of formal vector fields. Namely, *every non-zero formal vector field is conjugate with a vector field of the form*

$$w = (\pm x^r + \alpha x^{2r-1})\frac{\partial}{\partial x} \tag{2.5.3}$$

where the integer $r \geq 1$ *and real number* α *are invariant with respect to the conjugacy.*

A proof of local smooth conjugacy for flows is not only easier than for diffeomorphisms, but may be carried out in analytic situations as well.

We remind the reader that given a C^∞-function $v(x)$ we denote by $\hat{v}(x)$ its formal Taylor series at $x = 0$. We say that vector fields v and w are *formally conjugate at* $x = 0$ if $\hat{v}$ and $\hat{w}$ are conjugate as formal vector fields.

Theorem 2.17. *If v is a C^∞ (or analytic) vector field, non-flat at a singular point $x = 0$, and a vector field w is formally conjugate with v, then v and w are locally C^∞ (analytically) conjugate.*

Proof. We can assume that w is normal form (2.5.3) with $r \geq 2$ and that

$$v(x) = w(x) + x^{2r}h(x),$$

where h is a C^∞ (analytic) function. If a conjugating transformation has the form $\Phi(x) = x + x^r\varphi(x)$, then φ satisfies the equation

$$(\pm x^r + \alpha x^{2r-1})(1 + rx^{r-1}\varphi(x) + x^r\varphi'(x))$$

$$= \pm(x + x^r\varphi(x))^r + \alpha(x + x^r\varphi(x))^{2r-1} + (x + x^r\varphi(x))^{2r}h(x + x^r\varphi(x)).$$

After obvious transformations we arrive at the equation

$$(1 \pm \alpha x^{r-1})\varphi'(x) = \tilde{h}(x, \varphi(x)) \tag{2.5.4}$$

with a function $\tilde{h}(x, y)$ belonging to the same class as v and w. Since $1 \pm \alpha x^{r-1} \neq 0$ in a neighborhood of $x = 0$, the latter differential equation has a local C^∞ (analytic) solution $\varphi(x)$, and every local diffeomorphism $\Phi(x) = x + x^r\varphi(x)$ with such a solution $\varphi(x)$ conjugates v and w. □

Corollary 2.14. *Every local C^∞ (or analytic) vector field, non-flat at a singular point $x = 0$, may be locally reduced to normal form* (2.5.3) *using a C^∞ (analytic) conjugacy.*

2.5.2 Flows on the real line with hyperbolic fixed points

Two C^k-vector fields locally conjugate at a unique joint fixed point may not be globally conjugate on $\mathbb{R}$: one of them may be complete (for example, a linear field from Theorem 2.16) while the other is not. We will show now that this difference is the only obstacle for global conjugacy. For this purpose, let F^t and G^t be two flows generated by complete C^k-vector fields locally conjugate at the unique fixed point $x = 0$. Then there exists a local diffeomorphism Φ_0 in a neighborhood V of $x = 0$ such that

$$\Phi_0 F^t(x) = G^t\Phi_0(x), \quad x \in V, \ |t| < t_0.$$

Since $x = 0$ is the unique singular point of both vector fields, it is either attracting or repelling point for diffeomorphisms $F^t(x)$ and $G^t(x)$ on every semi-axis $x \in \mathbb{R}_\pm$, and $t \in \mathbb{R}_\pm$. To be definite we assume that $x = 0$ is attracting for F^t and G^t, $t > 0$. Similar to Section 2.3 it means that

$$\lim_{t\to+\infty} F^t(x) = \lim_{t\to+\infty} G^t(x) = x, \quad x \in \mathbb{R}.$$

Then for every $x \in \mathbb{R}$ the expression $G^{-t}\Phi_0 F^t(x)$ is defined for all sufficiently large t and does not depend on t. Hence we can define

$$\Phi(x) = \lim_{t\to+\infty} G^{-t}\Phi_0 F^t(x), \quad x \in \mathbb{R},$$

and easily check that Φ conjugates F^t and G^t globally.

If, in particular,

$$F^t(x) = e^{\alpha t}x + f(t,x), \quad f(t,0) = f'_x(t,0) = 0, \quad \alpha \neq 0,$$

is a C^k-flow with the unique fixed point $x = 0$ and $k \geq 2$, then according to Theorem 2.16 the vector field

$$v(x) = \frac{d}{dt} F^t(x)\,|_{t=0} \tag{2.5.5}$$

corresponding to F^t is locally C^k-conjugate with a linear vector field $v_\alpha(x) = \alpha x$. Hence F^t is globally C^k-conjugate with the linear flow $\Lambda^t(x) = e^{\alpha t}x$. In the case $k = 1$, similar to the discrete situation, some additional invariants arise. To describe them, let $\alpha < 0$ and let $\Psi_\pm$ be straightening C^k-diffeomorphisms on $\mathbb{R}_\pm$ for the field (2.5.5). Then

$$\Psi_\pm(F^t(s)) = \Psi_\pm(s) + t, \quad t \in \mathbb{R},\ s \in \mathbb{R}_\pm,$$

and

$$\lim_{u\to+\infty} \frac{(\Psi_\pm^{-1})'(u + t)}{(\Psi_\pm^{-1})'(u)} = e^{\alpha t}.$$

Hence

$$\Psi_\pm^{-1}(u) = \pm \int_u^{+\infty} e^{\alpha s}\theta_\pm(s)ds, \quad u > 0,$$

where $\theta_\pm(s) > 0$ are continuous functions satisfying the condition

$$\lim_{s\to+\infty} \frac{\theta_\pm(s + t)}{\theta_\pm(s)} = 1, \quad t \in \mathbb{R}. \tag{2.5.6}$$

Similar to functional coordinates of an individual C^1-diffeomorphism introduced in Section 2.3.2, a pair $\theta_\pm(t)$ may be accepted as *functional coordinates* of F^t.

If $\tilde{\Psi}_\pm$ is another pair of straightening diffeomorphisms for the same flow F^t, then $\tilde{\Psi}_\pm = H_\pm\,\Psi_\pm$ with some C^1-diffeomorphisms $H_\pm : \mathbb{R} \to \mathbb{R}$ commuting with the standard flow $F_0^t(s) = s + t$. Every such diffeomorphism has the form $H(s) = s + c$ where c is a constant. Therefore for the corresponding functional coordinates $\theta_\pm(t)$ and $\tilde{\theta}_\pm(t)$ we have

$$\tilde{\theta}_\pm(s) = \theta_\pm(s - c_\pm)\, e^{\alpha c_\pm}.$$

Let now $\tilde{F}^t$ be another C^1-flow with straightening diffeomorphisms $\tilde{\Psi}_\pm(s)$ and $\tilde{F}^t = \Phi F^t \Phi^{-1}$. Then the diffeomorphisms $H_\pm = \tilde{\Psi}_\pm \Phi \Psi_\pm^{-1}$ commute with the standard flow F_0^t, and hence $H_\pm(u) = u + d_\pm$.

Theorem 2.18. *Let*

$$F^t(x) = e^{\alpha t}x + f(t,x), \ \tilde{F}^t(x) = e^{\alpha t}x + \tilde{f}(t,x),$$

$$f(t,0) = \tilde{f}(t,0) = f'_x(t,0) = \tilde{f}'_x(t,0) = 0, \quad \alpha < 0,$$

be two C^1-flows with functional coordinates $\theta_\pm$ and $\tilde{\theta}_\pm$, respectively. Then F^t and $\tilde{F}^t$ are locally C^1-conjugate at $x = 0$ if and only if the limits

$$\lim_{s\to+\infty}\frac{\tilde{\theta}_+(s)}{\theta_+(s)}, \qquad \lim_{s\to+\infty}\frac{\tilde{\theta}_-(s)}{\theta_-(s)}$$

exist and are different from zero.

We omit the proof, since it repeats arguments used to prove Theorem 2.8.

Corollary 2.15. *A C^1-flow $F^t(x) = e^{\alpha t}x + f(t,x)$, $f(t,0) = f'_x(t,0) = 0$, with functional coordinates $\theta_\pm(s)$ is locally C^1-conjugate with the linear flow $\Lambda^t(x) = e^{\alpha t}x$ if and only if $\theta_\pm(s) = c_\pm + o(1)$, $s \to +\infty$, with $c_\pm \neq 0$.*

Let us now consider flows F^t with several fixed points.

Let $x_1 < \dots < x_n$, $n \geq 2$, be fixed points of a C^k-flow F^t and let

$$F^t(x) = x_i + e^{\alpha_i t}(x - x_i) + o(x - x_i), \quad x \to x_i, \quad i = 1, \dots, n.$$

Assume $\alpha_{2j-1} < 0 < \alpha_{2j}$, $j = 1, 2, \dots$, and set

$$U_1 = (-\infty, x_2); \ U_j = (x_{j-1}, x_{j+1}), \quad 2 \leq j \leq n-1; \quad U_n = (x_{n-1}, +\infty).$$

If $k \geq 2$, then it follows immediately from Theorem 2.16 that there exist C^k-diffeomorphisms $\Psi_i : U_i \to \mathbb{R}$, $\Psi_i(x_i) = 0$, such that

$$F^t|_{U_i} = \Psi_i^{-1}\Lambda_i^t\Psi_i, \ \Lambda_i^t(x) = e^{\alpha_i t}x, \ i = 1, 2, \dots, n.$$

In other words, the flow F^t is glued from the linear flows

$$\Lambda_i^t(x) = e^{\alpha_i t}x.$$

The *gluing diffeomorphisms* $G_i = \Psi_{i+1}\Psi_i^{-1}$, $i = 1, \dots, n-1$, map $\mathbb{R}_+$ onto $\mathbb{R}_-$ and satisfy conditions

$$G_i(e^{\alpha_i t}x) = e^{\alpha_{i+1}t}G_i(x), \quad t \in \mathbb{R}, \ x \in \mathbb{R}_+.$$

For $t = 1$ these are the same equalities as (2.4.6), but since t is an arbitrary number, instead of (2.4.7) we obtain

$$G_i(x) = -c_i x^{\nu_i}, \quad c_i > 0, \ \nu_i = \frac{\alpha_{i+1}}{\alpha_i}.$$

Denote by $C_0(k, \alpha_1, \ldots, \alpha_n)$ the set of all C^k-flows F^t on $\mathbb{R}$ with n fixed points $y_1 < \cdots < y_n$ such that

$$F^t(x) = y_i + e^{\alpha_i t}(x - y_i) + o(x - y_i), \quad x \to y_i, \quad i = 1, \ldots, n.$$

As before we assume that for $k = 1$ all restrictions $F^1|_{U_i}$, $i = 1, 2, \ldots, n$, are smoothly conjugate with their linear parts and hence $F^t|_{U_i}$, $i = 1, 2, \ldots, n$, are conjugate with corresponding linear flows.

Theorem 2.19. *All flows* $F^t \in C_0(k, \alpha_1 \ldots \alpha_n)$ *are* C^k*-conjugate.*

Proof. Let us fix a C^k-flow $F^t \in C_0(k, \alpha_1 \ldots \alpha_n)$ with the collection of gluing diffeomorphisms $G_i(x) = -x^{\nu_i}$ and assume $\tilde{F}^t \in C_0(k, \alpha_1 \ldots \alpha_n)$ with a collection $\tilde{G}_i(x) = -c_i x^{\nu_i}$. Choose numbers $d_i > 0$ such that $d_{i+1} = c_i d_i^{\nu_i}$, $i = 1, 2, \ldots n-1$. Then

$$G_i(d_i x) = d_{i+1}\tilde{G}_i(x), \quad x \in \mathbb{R}_+, \ i = 1, \ldots n-1.$$

It follows from these equalities that the mapping

$$\Phi(x) = \Psi_i^{-1}(d_i \Psi_i(x)), \quad x \in U_i, \ i = 1, 2, \ldots, n$$

is a well-defined C^k-diffeomorphism of the real line. It is easy to check that Φ conjugates the flows F^t and $\tilde{F}^t$. □

Let now $P(x)$ be a polynomial such that $P(j) = 0$, $P'(j) = \alpha_j \neq 0$, $j = 1, \ldots, n$, and $P(x) \neq 0$ for $x \neq j$, $j = 1, \ldots, n$. Then

$$v_P(x) = \frac{P(x)}{1 + P^2(x)} \frac{\partial}{\partial x}$$

is a complete analytic vector field generating a $C(k, \alpha_1, \ldots, \alpha_n)$-flow F^t. According to Theorem 2.19 every complete C^k-vector field v on $\mathbb{R}$, $k \geq 2$, with n singular points $x_1, \ldots, x_n$ and $\alpha_j = v'(x_j)$, $j = 1, \ldots, n$, is C^k-conjugate with v_P. If, to be definite, $(-1)^{n-j}\alpha_j > 0$, then one of possible polynomials P is given by the formula

$$P(x) = \prod_{j=1}^{n}(x - j)((L_{n-1}(x))^2 + \delta^2)$$

where L_{n-1} is the polynomial of degree $n - 1$ satisfying interpolation data

$$L_{n-1}(j) = \sqrt{\frac{(-1)^{n-j}\alpha_j}{(j-1)!(n-j)!} - \delta^2}$$

with

$$\delta^2 = \min_{1 \leq j \leq n} \frac{(-1)^{n-j}\alpha_j}{(j-1)!(n-j)!}.$$

2.6 Embedding problem and iterative roots

Given a diffeomorphism F, the problem of embedding is to find a C^k-flow F^t such that

$$F = F^t|_{t=t_0}.$$

If such a flow exists, then F is said to be *C^k-embeddable.*

According to a notion of a local flow given in Section 2.5, the problem of *local embedding* is to find a local flow F^t such that $F(x) = F^{t_0}(x)$ in a neighborhood of a point x_0 where F is defined.

Assume that F is embeddable and $F = F^{t_0}$. Then the mapping $\Phi = F^s$, $s = \frac{t_0}{n}$ satisfies the equation

$$\Phi^n = \underbrace{\Phi \ldots \Phi}_{n\ times} = F. \qquad (2.6.1)$$

Every solution Φ of this equation is called an *iterative root of degree* n of the mapping F. Of course, the problem of existence of an iterative root is meaningful for an arbitrary mapping F, not necessarily invertible. We will use the normal forms and functional coordinates of mappings described above to study embedding problems for mappings with the simplest dynamical behavior.

2.6.1 Mappings without non-wandering points

Theorem 2.20. *A C^k-mapping $F : \mathbb{R} \to \mathbb{R}$ without fixed and critical points for $k \geq 1$ and strictly monotonic for $k = 0$ possesses a C^k-iterative root of arbitrary degree. If F is a surjective mapping and hence is a C^k-diffeomorphism, then it is C^k-embeddable.*

Proof. If F is surjective, then according to Theorem 2.2 there exists a C^k-diffeomorphism $\Psi : \mathbb{R} \to \mathbb{R}$ conjugating it with the standard shift and the relation

$$F^t(x) = \Psi^{-1}(\Psi(x) + t), \quad x \in \mathbb{R},$$

defines a C^k-flow with $F = F^t|_{t=1}$.

If now a mapping F is not surjective, then it is of type 2 as defined in Section 2.2.1. Since F^n is of the same type, then according to Theorem 2.2.1 there exists a C^k-diffeomorphism $\Psi : \mathbb{R} \to \mathbb{R}$ which conjugates F and F^n. The mapping $\Phi(x) = \Psi^{-1}F\Psi(x)$ is a C^k-iterative root of F of degree n. □

2.6.2 C^0-embedding

The following statement is true both for $M = \mathbb{R}$ and $M = \mathbb{T}$.

Theorem 2.21. *An orientation-preserving homeomorphism $F : M \to M$ whose non-wandering points are all fixed is embeddable in a topological flow.*

Proof. Let $U = (x', x'')$ be an adjacent interval for $\operatorname{Fix} F$ and $\Psi_U : U \to \mathbb{R}$ be a straightening homeomorphism for the restriction F_U :

$$\Psi_U F(x) = \Psi_U(x) + 1, \quad x \in U.$$

Setting

$$F^t(x) = \begin{cases} \Psi_U^{-1}(\Psi_U(x) + t), & x \in U, \\ x, & x \in \operatorname{Fix} F, \end{cases}$$

we obtain a topological flow F^t with $F = F^1$. □

2.6.3 Diffeomorphisms with a unique fixed point

Let us start with the local problems of embedding and existence of iterative roots in a neighborhood of a fixed point.

Theorem 2.22. *A local C^∞-diffeomorphism F with the isolated fixed point x_0 and with the formal Taylor series $\hat{F}$ at x_0 such that $\hat{F}^2(x) \neq x$ possesses a local C^∞-iterative root of arbitrary odd degree in a neighborhood of x_0. If, in addition, $F'(x_0) > 0$, then F is locally C^∞-embeddable.*

Proof. Without loss of generality we accept $x_0 = 0$ and consider the family of C^∞-mappings

$$G^t_{r,\beta}(x) = \frac{x}{(1 + rtx^r + \beta t x^{2r} + x^{2r} g_{r,\beta}(x,t))^{1/r}}$$

where $g_{r,\beta}(x,t)$ satisfies the equation

$$g(x,t) = \frac{\beta x^{-r}}{r} \ln(1 + rtx^r + \beta t x^{2r} + x^{2r} g(x,t)) - \beta t \tag{2.6.2}$$

coinciding with (2.3.10) for $t = 1$. Given an interval $[-a, a] \subset \mathbb{R}$, there exists a neighborhood U of $x_0 = 0$ such that (2.6.2) has a unique analytic solution $g(x,t)$ in $U \times [-a, a]$ with $g(0,t) = 0$. The straightforward calculation shows that for every bounded interval $I \in \mathbb{R}$ there exists $\epsilon > 0$ such that

$$W_- G^t_{r,\beta}(x) = W_-(x) + t, \qquad -\epsilon < x < 0, \qquad t \in I,$$

$$W_+ G^t_{r,\beta}(x) = W_+(x) + t, \qquad 0 < x < \epsilon, \qquad t \in I,$$

where

$$W_+(x) = \frac{x^{-r}}{r} + \frac{\beta}{r} \ln x, \;\; x > 0; \;\; W_-(x) = \frac{x^{-r}}{r} + \frac{\beta}{r} \ln |x|, \;\; x < 0,$$

are the same as in Section 2.3.2. It means that the family $G^t_{r,\beta}(x)$ is a local analytic flow with the straightening diffeomorphisms $W_\pm$.

If now $F'(0) > 0$, then according to Theorem 2.7 the mapping F is locally C^∞-conjugate with either $G^1_{r,\beta}$ or $G^{-1}_{r,\beta}$ where r and β are some appropriate parameters. Hence F is locally C^∞-embeddable.

In the case $F'(0) < 0$ the mapping F is reduced to the formal normal form IV. According to the same Theorem 2.7 it is locally C^∞-conjugate with either $G(x) = G^1_{2s,\beta}(-x)$ or $G(x) = G^{-1}_{2s,\beta}(-x)$. Since $r = 2s$ is even, the function $g(x,t)$ from (2.6.2) is even with respect to x and $G^t_{2s,\beta}(-x) = -G^t_{2s,\beta}(x)$. For every odd n the local C^∞-mapping $\Phi(x) = G^{\pm 1/n}_{2s,\beta}(-x)$ satisfies relation $\Phi^n(x) = G^{\pm 1}_{2s,\beta}(-x)$ and is an iterative root of $G(x)$. Hence F possesses a local C^∞-iterative root of arbitrary odd degree. □

Corollary 2.16. *Let a diffeomorphism $F : \mathbb{R} \to \mathbb{R}$ with a unique fixed point x_0 be locally C^k-conjugate at x_0 with a C^∞-diffeomorphism G such that $\hat{G}^2(x) \neq x$. Then F possesses a global C^k-iterative root of every odd degree. If, in addition, $F'(x_0) > 0$, then F is C^k-embeddable.*

Proof. We have to check that a local flow F^t, $F^t|_{t=1} = F$, and a local solution Φ of (2.6.1), existing according to the previous theorem, may be continued onto the entire real line. This is possible by virtue of the commutativity conditions

$$F^t F = F F^t, \quad \Phi F = F \Phi.$$

□

In particular, if $F : \mathbb{R} \to \mathbb{R}$ is an orientation-preserving C^k-diffeomorphism, $k \geq 2$, with a unique hyperbolic fixed point then it is C^k-embeddable.

In contrast to the case $k \geq 2$, a C^1-diffeomorphism with a unique hyperbolic fixed point may not be C^1-embeddable. A condition for F to be embeddable may be stated in terms of functional coordinates $\theta_\pm$ introduced in Section 2.3.2.

Theorem 2.23. *An orientation-preserving C^1-diffeomorphism $F : \mathbb{R} \to \mathbb{R}$ with a unique hyperbolic fixed point x_0 and functional coordinates $\theta_\pm$ at x_0 is C^1-embeddable if and only if there exist C^1-diffeomorphisms $H_\pm$ commuting with the standard shift such that*

$$\lim_{z \to +\infty} \frac{\theta_\pm(H_\pm(z+s))H'_\pm(z+s)e^{H_\pm(z+s)\ln\lambda}}{\theta_\pm(H_\pm(z))H'_\pm(z)e^{H_\pm(z)\ln\lambda}} = e^{s\ln\lambda}, \; s \in \mathbb{R}, \tag{2.6.3}$$

with $\lambda = F'(x_0)$.

Proof. Let F^t be a C^1-flow and

$$F(x) = F^t(x)|_{t=1}.$$

Since F commutes with F^t, $t \in \mathbb{R}$, the point $F^t(x_0)$ is fixed for F and because of its uniqueness coincides with x_0 : $F^t(x_0) = x_0$. We can assume $x_0 = 0$ and

$$F^t(x) = e^{\alpha t}x + f(t,x), \quad \alpha = \ln F'(x_0), \quad f(t,0) = f'_x(t,0) = 0.$$

According to Section 2.5.2 functional coordinates $\tilde\theta_\pm(z)$ of F^t at $x=0$ satisfy relations

$$\lim_{z\to+\infty}\frac{\tilde\theta_\pm(z+s)}{\tilde\theta_\pm(z)} = 1, \quad s\in\mathbb{R}.$$

Of course, $\tilde\theta_\pm(z)$ are functional coordinates of F at $x=0$ as well. As we noted in Section 2.3.2, there exist diffeomorphisms $H_\pm$ commuting with the standard shift such that

$$\tilde\theta_\pm(z) = \theta(H_\pm(z))H'_\pm(z)e^{(H_\pm(z)-z)\ln\lambda}$$

and (2.6.3) holds.

Conversely, let (2.6.3) hold. Then

$$\tilde\theta_\pm(z) = \theta_\pm(H_\pm(z))H'_\pm(z)e^{(H_\pm(z)-z)\ln\lambda}$$

are functional coordinates of a C^1-flow

$$F^t(x) = e^{\alpha t}x + f(t,x), \quad \alpha = \ln F'(x_0).$$

The diffeomorphism F is C^1-conjugate with $F^t|_{t=1}$. Hence F is C^1-embedded in F^t. □

A diffeomorphism F with a unique hyperbolic fixed point x_0, $F'(x_0)=\lambda\in(0,1)$ and functional coordinates

$$\theta_+(z) = z^\delta, \quad \theta_-(z) = 1, \quad \delta\neq 0,$$

is an example of a C^1-embeddable diffeomorphism which is not smoothly conjugate with its linear part as follows from Corollary 2.6.

Example 2.10. Let F be a diffeomorphism with a unique hyperbolic fixed point x_0, $F'(x_0) = \lambda\in(0,1)$, and with functional coordinates

$$\theta_+(z) = \exp((\sin 2\pi z)\ln z), \quad \theta_-(z) = 1.$$

We will show that F is not smoothly embeddable. To this end we will check that for any diffeomorphism H commuting with the standard shift, relation (2.6.3) does not hold for some $s\in\mathbb{R}$. Indeed, assume that $2H(0)$ is an integer and choose $t_0\in\mathbb{R}$ such that $4H(t_0) = 1$. Since $H(k+s) = k+H(s)$ for every integer k, then

$$\frac{\theta_+(H(k+t_0))H'(k+t_0)}{\theta_+(H(k))H'(k)} = \left(k+\frac{1}{4}\right)\frac{H'(t_0)}{H'(0)}, \quad k\in\mathbb{Z},$$

and the limit in (2.6.3) does not exist for $s=t_0$. If $2H(0)$ is not an integer, we choose t_0 such that $H(t_0) = 0$ and obtain

$$\frac{\theta_+(H(k+t_0))H'(k+t_0)}{\theta_+(H(k))H'(k)} = \frac{H'(t_0)}{e^{\delta\ln(k+H(0))}H'(0)}, \quad k\in\mathbb{Z}$$

with $\delta = \sin 2\pi H(0)\neq 0$. In any case (2.6.3) does not hold and hence a C^1-diffeomorphism F with the functional coordinates $\theta_\pm(x)$ is not C^1-embeddable.

In the analytic situation an embedding condition coincides with the condition of conjugacy with the normal form $G_{r,\beta}(x)$.

Theorem 2.24. *A local analytic diffeomorphism* $F(x) = G_{r,\beta}(x) + O(x^k)$, $k \geq 2r + 2$, *is locally analytically embeddable if and only if* (2.3.23) *holds.*

Proof. If (2.3.23) holds, then according to Corollary 2.19 the diffeomorphism F is analytically conjugate with $G_{r,\beta}(x)$ and the latter is embeddable in the analytic flow $G^t_{r,\beta}(x)$. Conversely, let $F = F^t|_{t=1}$ where F^t is a local analytic flow. It follows from Theorem 2.17 that F^t is analytically conjugate with $G^t_{r,\beta}$ and hence F is conjugate with $G_{r,\beta} = G^1_{r,\beta}$. Again using Corollary 2.19 we conclude that (2.3.23) holds. □

Comparing Theorem 2.24 and Example 2.8 we find that the local analytic diffeomorphism

$$F_\delta(x) = \frac{x}{1 + x + \delta x^3}$$

with small δ is not analytically embeddable.

2.6.4 Diffeomorphisms with several fixed points

If F has several fixed points, then the existence of a local embedding at every such point is not sufficient for the existence of a global embedding. Similar to Section 2.4 some obstacles arise related to the gluing diffeomorphisms.

Theorem 2.25. *Let* $F \in C(k, \lambda_1, \ldots, \lambda_q)$, $q \geq 2$, *be a diffeomorphism with gluing diffeomorphisms*

$$G_j(x) = -x^{\alpha_j}\tau_j\left(\frac{\ln x}{\ln \lambda_j}\right), \quad \alpha_j = \frac{\ln \lambda_{j+1}}{\ln \lambda_j}, \; j = 1, \ldots q-1. \tag{2.6.4}$$

Then F *possesses a smooth iterative root of degree* n *if and only if the functions* $\tau_j(t)$ *are* n^{-1}*-periodic, and* F *is smoothly embeddable if and only if all these functions are constant.*

Proof. Let Φ be a smooth solution of (2.6.1) and

$$\mathrm{Fix}\, F = \{x_1, \ldots, x_q\}.$$

It is easy to check that

$$\mathrm{Fix}\, \Phi = \{x_1, \ldots, x_q\}, \quad \Phi'(x_j) = \sqrt[n]{\lambda_j}\,.$$

Let us show that $\Phi \in C(1, \sqrt[n]{\lambda_1}, \ldots, \sqrt[n]{\lambda_q})$, which means that for every $j = 1, \ldots, q$, the restriction $\Phi|_{U_j}$ is C^1-conjugate with its linear part $\sqrt[n]{\lambda_j}x$. The latter follows immediately from Theorem 2.5 if $\Phi \in C^p$, $p \geq 2$. For $p \geq 1$ let $\Psi_j : U_j \to \mathbb{R}$ be a diffeomorphism conjugating $F|_{U_j}$ with its linear part, i.e.,

$$\Psi_j F \Psi_j^{-1}(z) = \lambda_j z, \quad z \in \mathbb{R}.$$

Then the diffeomorphism $\tilde{\Phi}_j = \Psi_j \Phi \Psi_j^{-1} : U_j \to U_j$ satisfies $\tilde{\Phi}_j^n(z) = \lambda_j z$. Substituting here $\tilde{\Phi}_j(z)$ instead of z we obtain

$$\lambda_j \tilde{\Phi}_j(z) = \tilde{\Phi}_j(\lambda_j z), \quad z \in \mathbb{R},$$

which means that $\tilde{\Phi}_j$ commutes with the hyperbolic linear mapping. Being a C^1-diffeomorphism, the mapping $\tilde{\Phi}_j$ is linear. Therefore $\tilde{\Phi}_j(z) = \sqrt[n]{\lambda_j} z$, and $\Psi_j \Phi \Psi_j^{-1}(z) = \sqrt[n]{\lambda_j} z$, $z \in U_j$.

The gluing diffeomorphisms $\tilde{G}_j$ for Φ have the form

$$\tilde{G}_j(x) = -x^{\alpha_j} \tilde{\tau}_j \left(\frac{n \ln x}{\ln \lambda_j} \right), \quad x \in \mathbb{R}_+, \quad j = 1, \ldots, q,$$

with smooth 1-periodic functions $\tilde{\tau}_j$, and it follows from (2.6.1) that they are gluing diffeomorphisms for F as well. Hence the collection $\{ \tilde{G}_j \}$ is equivalent to the collection $\{ G_j \}$, and (2.4.8) implies

$$\tau_j(t) = b_j \tilde{\tau}_j(nt + c_j), \quad b_j > 0, \quad c_j \in \mathbb{R}.$$

Therefore τ_j, $j = 1, \ldots, q$, are n^{-1}-periodic functions.

Conversely, let $\{\Psi_j\}$ be a collection of diffeomorphisms conjugating the restrictions $F|_{U_j}$ with their linear parts at the corresponding fixed points and let the collection of gluing diffeomorphisms $G_j = \Psi_{j+1} \Psi_j^{-1}$ be of the form (2.6.4) where τ_j are n^{-1}-periodic functions. The function

$$\Phi(x) = \Psi_j^{-1}(\sqrt[n]{\lambda_j} \Psi_j(x)), \quad x \in U_j,$$

satisfies $\Phi^n = F$ on every set U_j and it remains to check that it is a well-defined C^k-diffeomorphism of M. It means that the identities

$$\Psi_j^{-1}(\sqrt[n]{\lambda_j} \Psi_j(x)) = \Psi_{j+1}^{-1}(\sqrt[n]{\lambda_{j+1}} \Psi_{j+1}(x)), \; x \in U_j \bigcap U_{j+1}, \; j = 1, \ldots, q,$$

hold, which is equivalent to

$$G_j(\sqrt[n]{\lambda_j} x) = \sqrt[n]{\lambda_{j+1}} G_j(x), \qquad x \in \mathbb{R}_+.$$

The latter follows from the assumption that τ_j is an n^{-1}-periodic function.

Assume now that a diffeomorphism F is C^1-embedded. Then it possesses smooth iterative roots of arbitrary degree n and the functions $\tau_j(t)$, being n^{-1}-periodic for every $n = 1, 2, \ldots$, are constant. On the other hand, for $\tau_j(t) =$const we have

$$G_j(x) = -c_j x^{\alpha_j}.$$

For any flow $F^t \in C_0(k, \nu_1, \ldots, \nu_q)$, $\nu_j = \ln \lambda_j$, the diffeomorphism F is C^k-conjugate with F^1 and hence it is embeddable in a C^k-flow. □

Corollary 2.17. *If a diffeomorphism $F \in C(k, \lambda_1, \dots, \lambda_q)$ possesses a C^1-iterative root, then it possesses a C^k-iterative root of the same degree. If F is C^1-embedded then it is C^k-embedded as well.*

It follows from Theorem 2.25 that an analytic diffeomorphism $F \in C(\aleph, \lambda_1, \dots, \lambda_q)$ with

$$\tau_j(t) = 1 + \varepsilon \sin 2\pi t, \quad j = 1, \dots, q-1,$$

and sufficiently small $\epsilon \neq 0$ has no smooth iterative roots of degree $n > 1$. At last, given an integer n, a diffeomorphism $F \in C(k, \lambda_1, \dots, \lambda_1)$ with

$$\tau_j(t) = 1 + \varepsilon \sin 2\pi n t, \quad j = 1, \dots, q-1,$$

has a C^k-iterative root of degree m if and only if m is a divisor of n.

2.7 Comments and references

2.1. The notion of non-wandering points is widely known in dynamics ([66]). Wandering subsets in the form of Definition 2.1 first appeared in [23] in connection with the solvability of Abel equations on arbitrary topological spaces.

There are many publications dedicated to the classification of mappings $F : \mathbb{T} \to \mathbb{T}$ (see [66]). The essential difference between the real line and the circle is caused by the presence of non-wandering points for every homeomorphism $F : \mathbb{T} \to \mathbb{T}$. Poincaré introduced a topological invariant $\tau(F)$ which is called the *rotation number*. If $\tau(F)$ is an irrational number and $F \in C^2$, then according to the *Danjoy Theorem*, F is topologically conjugate with the rotation by $\tau(F)$. A smooth conjugacy with the rotation was studied by Herman [47]. If τ is rational, then all non-wandering points are periodic. A detailed study of mappings of the unit circle into itself may be found in [7].

2.2. Theorems 2.1 and 2.2 for smooth mappings F are well known [53]. As far as we know the analytic version of Theorem 2.2 was first published in [30] where functional equations were studied in a neighborhood of non-fixed points. Let us sketch a proof of Theorem 2.2 based on the uniqueness of a smooth structure which was already used in the proof of Theorems 2.12 and 2.13.

Denote by $Z = \mathbb{R}/F$ the orbit manifold of a C^k-diffeomorphism $F : \mathbb{R} \to \mathbb{R}$ without fixed points. The diffeomorphism F induces on Z a structure of a C^k-manifold homeomorphic to the unit circle $\mathbb{T}$ which, in its turn, is the orbit manifold induced by the standard shift. Since a C^k-structure is unique, there exists a C^k-diffeomorphism $\Psi : Z \to \mathbb{T}$. If $\Phi : \mathbb{R} \to \mathbb{R}$ is a C^k-lifting [49] of Ψ, then for every $x \in \mathbb{R}$ the points $\Phi(x)$ and $\Phi(F(x))$ belong to the same orbit of the standard shift. Therefore $\Phi(F(x)) = \Phi(x) + n(x), n(x) \in \mathbb{Z}$, and since $n(x)$ is a continuous function, it is identically equal to a constant n. The C^k-diffeomorphism $\tilde{\Phi} = n^{-1}\Phi$ conjugates F with the standard shift. Note that the above arguments are valid for all C^k-classes, including analytic functions.

Theorem 2.3 is a one-dimensional version of a general result from [23] dealing with the cohomological equation on arbitrary topological spaces X. In particular, it was proved in [23] that if $F : X \to X$ is an injective continuous mapping of a locally compact space X countable at infinity, then the solvability of the Abel equation in continuous functions implies the solvability of the cohomological equation with an arbitrary continuous function γ. Example 2.2 shows that this statement fails for non-injective mappings, and the problem arises to find conditions for $F : X \to X$ to guarantee the solvability of the Abel equation. The inverse to Proposition 2.1 looks to be plausible:

If all compact subsets are wandering for F, then the Abel equation has a continuous solution.

Another problem is connected to Theorem 2.4 and Example 2.3:

What conditions guarantee smooth solvability of the Abel and cohomological equations ?

The study of the cohomological equation on $\mathbb{T}$ was initiated by Anosov [1]. Many results which followed may be found in [51]. Due to the presence of non-wandering points, the Abel equation has no bounded solutions. If F is a rotation by an irrational number, then the condition

$$\int_0^{2\pi} \gamma(e^{it})dt = 0$$

is necessary for the cohomological equation (2.2.6) to have a measurable solution φ. The necessary conditions of such a type appear for arbitrary mappings, but usually they are not sufficient for the solvability, since the image of the operator $L\varphi(x) = \varphi(F(x)) - \varphi(x)$ may not be closed in corresponding spaces. We will discuss this phenomenon in Chapter 5.

2.3. Theorem 2.5 is a one-dimensional version of the *Sternberg Theorem* [72] for smooth mappings and of the *Poincaré Theorem* [67] for analytic mappings. Formal normal forms III and IV were found in [62], see also [53]. Formal normal forms for multi-dimensional mappings were constructed in [15]. Theorem 2.7 was obtained independently in [13] and [73]. The approach to smooth classification based on the straightening on a semi-axis is presented here for the first time. Another proof of Theorem 2.7 based on solving the Schröder equation will be given in Theorem 3.3, Chapter 3 below. The Ecalle–Voronin moduli for analytic classification have arrived comparatively recently, see [41, 75, 65], although examples demonstrating that the formal conjugacy does not imply the analytic one were known earlier, see [53].

2.4. Functional invariants of the global smooth classifications were found in [61] for the interval $[0, 1]$, in [17] for the unit circle and in [18] for the real line, see also [8, 51]. Multi-dimensional generalizations and further effects were described in [16]. In particular, functional moduli of topological conjugacy on a two-dimensional torus were constructed in [16] by the same gluing method.

2.5. The proof of Theorem 2.17 is based on a reduction of the conjugacy problem to the solvability of non-singular differential equation (2.5.4). It may be shown that even a general singular equation $x^r\varphi'(x) = f(x, \varphi(x))$, $f \in C^\infty$, is locally C^∞-solvable whenever it is formally solvable at $x = 0$. The simple examples show that this is not true in the analytic situation. For instance, the equation

$$x^2\varphi'(x) = \varphi(x) - x$$

studied by Euler back in 1756 has the unique formal solution

$$\hat{\varphi}(x) = \sum_{n=1}^{\infty}(n-1)!x^n$$

which diverges if $x \neq 0$.

Contrary to the case of the real line, some numerical invariants arise in smooth classification of flows with non-degenerate singular points on the unit circle. These invariants were found in [34, 35, 48].

2.6. Various results on iterative roots may be found in [53, 54, 12] and in references contained there. More detailed exposition of results on embedding and iterative roots in the analytic situation may be found in [65]. In particular, conditions of existence of analytic iterative roots may be stated in terms of the Ecalle–Voronin moduli $h_j(t)$ similar to Theorem 2.25.

According to Theorem 2.20 for every integer $n \geq 2$ there exists a real-analytic function F satisfying equation

$$\underbrace{\Phi \ldots \Phi}_{n\ times}(x) = e^x.$$

It follows from [11] that this equation has no entire solutions. Yu. Lyubich informed us that there are no non-trivial representations $e^x = f(g(x))$ with entire functions f and g.

Chapter 3

Generalized Abel Equation

Generalized Abel equations have the form

$$\varphi(F(x)) = g(x, \varphi(x)) \tag{3.0.1}$$

where $F : M \to M$ is a given mapping, $g(x,y)$ is a given function of $x \in M, y \in \mathbb{R}$ and $\varphi(x)$ is a solution. The Abel, Schröder and cohomological equations are particular cases of (3.0.1). Our aim is to find solvability conditions and, if possible, to describe the set of solutions of equation (3.0.1) in terms of $F(x)$ and $g(x,y)$.

There exists a natural correspondence between generalized Abel equations (3.0.1) and mappings

$$G(x,y) = (F(x), g(x,y)).$$

In the simplest cases $G(x,y) = (x+1, y)$, $G(x,y) = (x+1, -y)$ the corresponding equations take on the form

$$\varphi(x+1) = \varphi(x), \quad \varphi(x+1) = -\varphi(x),$$

and their solutions are either 1-periodic or 1-anti-periodic functions, respectively.

To investigate equation (3.0.1) let

$$\Phi(x,y) = (H(x), T(x,y)) \tag{3.0.2}$$

be an invertible transformation of the manifold $M \times \mathbb{R}$ into itself and let $\tilde{G}(x,y) = \Phi^{-1}G\Phi(x,y)$. Then $\tilde{G}(x,y) = (\tilde{F}(x), \tilde{g}(x,y))$ with

$$\tilde{F} = H^{-1}FH, \quad T(\tilde{F}(x), \tilde{g}(x,y)) = g(H(x), T(x,y)). \tag{3.0.3}$$

Let us assume that $\tilde{\varphi}$ is a solution of

$$\tilde{\varphi}(\tilde{F}(x)) = \tilde{g}(x, \tilde{\varphi}(x)) \tag{3.0.4}$$

and set

$$\varphi(x) = T(H^{-1}(x), \tilde{\varphi}(H^{-1}(x))). \tag{3.0.5}$$

Then

$$\begin{aligned}\varphi(F(x)) &= T(H^{-1}(F(x)), \tilde{\varphi}(H^{-1}(F(x)))) \\ &= T(\tilde{F}(H^{-1}(x)), \tilde{\varphi}(\tilde{F}(H^{-1}(x)))) \\ &= T(\tilde{F}(H^{-1}(x)), \tilde{g}(H^{-1}(x), \tilde{\varphi}(H^{-1}(x)))) \\ &= g(x, T(H^{-1}(x), \tilde{\varphi}(H^{-1}(x)))) = g(x, \varphi(x))\end{aligned}$$

which means that $\varphi(x)$ is a solution (3.0.1). Therefore relations (3.0.5) give one-to-one correspondences between solutions of (3.0.1) and (3.0.4) permitting us to use transformations (3.0.3) for simplifying equation (3.0.1).

3.1 Local solvability

First of all we clarify a notion of the local solvability.

Let $F(x)$ be a C^k-mapping defined in a neighborhood of a set $S \subset M$ and let $g(x,y)$ be a C^k-function defined on a neighborhood of $S \times V \subset M \times \mathbb{R}$ where $V \subset \mathbb{R}$.

Definition 3.1. Equation (3.0.1) is said to be *locally C^k-solvable* in a neighborhood of a set $S \subset M$ (at S, in short) if there exists a C^k-function φ defined in a connected neighborhood U of the union $S \bigcup F(S)$ and mapping a neighborhood W of S, $W \subset U$, in a neighborhood of $V \in \mathbb{R}$ such that (3.0.1) holds for all $x \in W$. The function φ is called a *local C^k-solution* of (3.0.1) at S.

In particular, if S is F-invariant, then a solution φ must be defined in a neighborhood U of S, not necessarily F-invariant, and (3.0.1) holds for x from a neighborhood W of S, $W \subset U$.

In this section we study local solutions of (3.0.1) at a one-point set S.

3.1.1 Local solvability in a neighborhood of a non-fixed point

Let us start with simple examples.

Example 3.1. If $F(x) = x^2 + 1$ and $\varphi(x)$ is a local solution of (3.0.1) at $x = 0$, then $g(x, \varphi(x))$ is an even function. This is impossible if $g(x,y) = g(x)$ does not depend on y and is not an even function in a neighborhood of $x = 0$. For such $g(x,y)$ there are no local continuous solutions of (3.0.1).

Example 3.2. In a non-analytic case the equation

$$\varphi(1 - x) = \varphi(x) + \gamma(x) \tag{3.1.1}$$

with $F(x) = 1 - x$, $x_0 = 0$, $F(x_0) = 1$, has a local C^k-solution for every C^k-function γ. Indeed, let

$$\varphi_0(x) = \begin{cases} \gamma(1 - x), & x \in (1-\varepsilon, 1+\varepsilon), \\ 0, & x \in (-\varepsilon, \varepsilon), \end{cases}$$

with $\varepsilon > 0$ sufficiently small. Every extension of the function φ_0 to a C^k-function φ in a neighborhood of $[0,1]$ is a local solution at $x_0 = 0$. On the other hand, if γ is an analytic function in a neighborhood of $[0,1]$ and $\gamma(2^{-1}) \neq 0$, then (3.0.1) has no local analytic solutions. Indeed, if φ is analytic in a neighborhood of $[0,1]$, then (3.1.1) holds for all $x \in [0,1]$ and $\gamma(2^{-1}) = 0$. Note that in this example F does not preserve the orientation.

We apply now the scheme described at the beginning of the present chapter to a study of local solutions of (3.0.1) at a non-fixed point.

Let $F(x)$ be a local C^k-mapping at x_0, $F(x_0) \neq x_0$, and let $g(x,y)$ be a C^k-function in a neighborhood of (x_0, y_0). Denote $x_0' = F(x_0)$, $y_0' = g(x_0, y_0)$ and assume $\tilde{F}(x)$ and $\tilde{g}(x,y)$ to be another pair of similar C^k-functions with corresponding points $\tilde{x}_0$, $\tilde{y}_0$, $\tilde{x}_0'$, $\tilde{y}_0' = \tilde{g}(\tilde{x}_0, \tilde{y}_0)$.

Let $\Phi(x,y)$ be a C^k-diffeomorphism of the form (3.0.2) defined in a connected neighborhood of the two-point set $(\tilde{x}_0, \tilde{y}_0) \bigcup (\tilde{x}_0', \tilde{y}_0')$ and conjugating $G(x,y)$ and $\tilde{G}(x,y)$ in a neighborhood of $(\tilde{x}_0, \tilde{y}_0)$. Then (3.0.5) defines a bijective correspondence between local C^k-solutions of (3.0.1) at x_0 and such solutions of (3.0.4) at $\tilde{x}_0$.

Theorem 3.1. *Let $F(x_0) \neq x_0$, $F'(x_0) > 0$ and let a mapping*

$$G(x,y) = (F(x), g(x,y))$$

be a local C^k-diffeomorphism of a neighborhood of the point $(x_0, y_0) \in \mathbb{R}^2$ onto a neighborhood of the point $(F(x_0), g(x_0, y_0))$. Then there exists a C^k-diffeomorphism $\Phi(x,y) = (H(x), T(x,y))$ of a neighborhood of the interval $[0,1] \subset \mathbb{R}^2$ onto a neighborhood of the two-point set

$$\{(x_0, y_0)\} \bigcup \{(F(x_0), g(x_0, y_0))\},$$

which conjugates $G(x,y)$ with the mapping

$$\tilde{G}(x,y) = (x + 1, \sigma y), \quad \sigma = \operatorname{sign} \frac{\partial g(x_0, y_0)}{\partial x},$$

in a neighborhood of the point (x_0, y_0).

In particular, the C^k-diffeomorphism H conjugates F with the standard shift in a neighborhood of x_0.

Using the connection between solutions of (3.0.1) and (3.0.4) described above we immediately obtain the following description of solutions to (3.0.1).

Corollary 3.1. *Let* $F(x)$ *and* $g(x,y)$ *satisfy conditions of Theorem* 3.1 *and* $\Phi(x,y) = (H(x), T(x,y))$ *be a mapping from its statement. Then every function* $\varphi(x)$ *defined by* (3.0.5) *with an arbitrary either* 1*-periodic function* $\tilde{\varphi}(x)$ *if* $g'_y(x_0, y_0) > 0$ *or* 1*-anti-periodic function* $\tilde{\varphi}(x)$ *if* $g'_y(x_0, y_0) < 0$, *is a local* C^k*-solution of* (3.0.1) *at* x_0. *Conversely, every such solution is representable by* (3.0.5).

Similar to Example 3.2, the condition $F'(x_0) > 0$ is redundant for the existence of solutions in a non-analytic case. Indeed, if we set

$$\varphi_0(x) = \begin{cases} y_0, & x \in (x_0 - \varepsilon, x_0 + \varepsilon), \\ g(F^{-1}(x), y_0), & x \in (F(x_0) - \varepsilon, F(x_0) + \varepsilon), \end{cases}$$

and extend φ_0 to a C^k-function on the interval $[x_0, F(x_0)]$, then for $x \in (x_0 - \varepsilon, x_0 + \varepsilon)$ we have $\varphi(F(x)) = g(x, y_0) = g(x, \varphi(x))$, and φ is a C^k-solution of (3.0.1). Of course, these arguments fail for analytic solutions.

Proof of Theorem 3.1 *for smooth functions.* To be definite, let us assume that $F(x_0) > x_0$ and $g'_y(x_0, y_0) > 0$.

To construct $H(x)$ we choose sufficiently small $\varepsilon > 0$ and define

$$H_0(x) = \begin{cases} x_0 + x, & x \in (-\varepsilon, \varepsilon), \\ F(x_0 + x - 1), & x \in (1 - \varepsilon, 1 + \varepsilon). \end{cases}$$

Since $H_0(0) = x_0 < F(x_0) = H_0(1)$ and $H'_0(x) > 0$, there exists a C^k-continuation $H(x)$ coinciding with $H_0(x)$ in neighborhoods of $x = 0$ and $x = 1$ and such that $H'(x) > 0$. This continuation satisfies

$$FH(x) = H(x + 1), \quad x \in (-\varepsilon, \varepsilon),$$

and therefore is a C^k-diffeomorphism conjugating F with the standard shift in a neighborhood of the origin.

To construct $T(x,y)$ we choose a C^k-function $\tau : \mathbb{R} \to [0,1]$ equal to 1 in a neighborhood V of $x = 1$ and vanishing outside a neighborhood W containing $\overline{V}$ and set

$$T(x,y) = (1 - \tau(x))(y + y_0) + \tau(x) g(H(x - 1), y + y_0).$$

If the neighborhood W is sufficiently small, then

$$\frac{\partial T(x,0)}{\partial y} = 1 - \tau(x) + \tau(x) \frac{\partial g(H(x - 1), y_0)}{\partial y} > 0$$

for $x \in [0,1]$ and the mapping $\Phi(x,y) = (H(x), T(x,y))$ is a C^k-diffeomorphism of a neighborhood of $[0,1] \subset \mathbb{R}^2$ to a neighborhood of $\{(x_0, y_0)\} \bigcup \{(F(x_0), g(x_0, y_0))\}$. The equality

$$g(H(x), T(x,y)) = T(x + 1, y), \quad x \in (-\varepsilon, \varepsilon),$$

follows, yielding (3.0.3). □

3.1.2 Proof of Theorem 3.1 for analytic functions

To prove Theorem 3.1 in the analytic case we use the *KAM method.* Its application requires some preparation.

As before we assume $F(x_0) > x_0$ and $g'_y(x_0, y_0) > 0$.

Lemma 3.1. *There exists an analytic diffeomorphism* $\Phi(x, y) = (H(x), T(x, y))$ *of a neighborhood of* $[0, 1] \subset \mathbb{R}^2$ *onto a neighborhood of* $\{(x_0, y_0)\} \bigcup \{(F(x_0), g(x_0, y_0))\}$, *such that*

$$\Phi^{-1}G\Phi(x, y) = (x + 1 + f(x), y + h(x, y))$$

where

$$f(0) = f'(0) = 0, \quad h(0,0) = 0, \quad \frac{\partial h(0,0)}{\partial x} = \frac{\partial h(0,0)}{\partial y} = 0. \tag{3.1.2}$$

Proof. First of all, we note that the transformation

$$\Phi_0(x, y) = (F(x_0) - x_0)x + x_0, (g(x_0, y_0) - y_0)x + y + y_0)$$

reduces G to the form

$$\Phi_0^{-1}G\Phi_0(x, y) = (cx + 1 + f_0(x), ay + h_0(x, y)), \tag{3.1.3}$$

where (x, y) belongs to a neighborhood of $(0, 0)$, and

$$c = F'(x_0), \quad a = \frac{\partial g(x_0, y_0)}{\partial y},$$

$$f_0(0) = f_0'(0) = 0, \quad h_0(0,0) = 0, \quad \frac{\partial h_0(0,0)}{\partial y} = 0.$$

Let now $\psi(x)$ be an increasing analytic function in a neighborhood of $[0, 1]$ such that

$$\psi(0) = 0, \quad \psi(1) = 1, \quad \psi'(0) = 1, \quad \psi'(1) = c.$$

Then the diffeomorphism

$$\Phi_1(x, y) = (\psi(x), e^{x \ln a} y)$$

transforms (3.1.3) to the form

$$\Phi_1^{-1}\Phi_0^{-1}G\Phi_0\Phi_1(x, y) = (x + 1 + f_1(x), y + h_1(x, y))$$

with

$$f_1(0) = f_1'(0) = 0, \quad h_1(0,0) = \frac{\partial h_1(0,0)}{\partial y} = 0.$$

If

$$\Phi_2(x,y) = (x, y + \frac{1}{2}bx(x-1)), \quad b = \frac{\partial h_1(0,0)}{\partial x},$$

then the transformation $\Phi = \Phi_0\Phi_1\Phi_2$ reduces G to the form (3.1.2). □

From now on we assume

$$G(x,y) = (x + 1 + f(x),\ y + g(x,y))$$

where

$$f(0) = f'(0) = 0, \quad g(0,0) = \frac{\partial g(0,0)}{\partial x} = \frac{\partial g(0,0)}{\partial y} = 0,$$

and apply the KAM method following the general scheme as described in [9].

Let $\Phi(x,y) = (x + \varphi(x),\ y + t(x,y))$ be an invertible transformation. Then for the transformed mapping

$$G_1(x,y) = \Phi G \Phi^{-1}(x,y) = (x + 1 + f_1(x),\ y + g_1(x,y))$$

we have

$$f_1(x + \varphi(x)) = \varphi(x + 1 + f(x)) - \varphi(x) + f(x)$$

and

$$\begin{aligned} &g_1(x + \varphi(x),\ y + t(x,y)) \\ &= t(x + 1 + f(x),\ y + g(x,y)) - t(x,y) + g(x,y). \end{aligned}$$

Assume that φ and t satisfy the cohomological equations

$$\varphi(x + 1) - \varphi(x) + f(x) = 0, \tag{3.1.4}$$

$$t(x + 1,\ y) - t(x,y) + g(x,y) = 0. \tag{3.1.5}$$

Then the functions

$$f_1(x + \varphi(x)) = \varphi(x + 1 + f(x)) - \varphi(x + 1) \tag{3.1.6}$$

and

$$g_1(x + \varphi(x),\ y + t(x,y)) = t(x + 1 + f(x),\ y + g(x,y)) - t(x + 1,\ y) \tag{3.1.7}$$

have a "smallness order" higher than the original functions f and g. Repeating the procedure, we obtain the converging sequences of transformations Ψ_n and mappings

$$G_n(x,y) = \Psi_n G \Psi_n^{-1}(x,y) = (x + 1 + f_n(x),\ y + g_n(x,y))$$

such that $f_n \to 0$, $g_n \to 0$ as $n \to \infty$. As a result we obtain

$$G_\infty(x,y) = \Psi_\infty G \Psi_\infty^{-1}(x,y) = (x + 1, y),$$

as stated in Theorem 3.1.

To carry out the above plan let us start with estimates of solutions of the cohomological equation.

Given a function γ analytic in a neighborhood of $(0,0) \in \mathbb{R}$, consider the equation

$$\psi(x+1,y) - \psi(x,y) = \gamma(x,y). \tag{3.1.8}$$

For $\rho > 0$ consider the following domains:

$$\begin{aligned}
\Delta(\rho) &= \{ z \in \mathbb{C}^1 : |z| < \rho \};\\
I(\rho) &= \{ z \in \mathbb{C}^1 : z = t + \xi,\ t \in [0,1],\ \xi \in \Delta(\rho) \};\\
\textstyle\prod(\rho) &= \{ (z_1, z_2) \in \mathbb{C}^2 : |z_i| < \rho \};\\
\Omega(\rho) &= \{ (z_1, z_2) \in \mathbb{C}^2 : z_1 = t + \xi,\ t \in [0,1],\ (\xi, z_2) \in \textstyle\prod(\rho) \}.
\end{aligned}$$

Lemma 3.2. *There exists a number $\nu > 0$ such that for every function $\gamma(z_1, z_2)$ analytic in a domain $\prod(\rho)$, $\rho \leq 2^{-1}$, and continuous in its closure $\overline{\prod}(\rho)$ and for every $\delta \leq 2^{-1}$, there exists a solution $\psi(z_1, z_2)$ of (3.1.8) analytic in $\Omega(\rho e^{-\frac{\delta}{4}})$, continuous in $\overline{\Omega}(\rho e^{-\frac{\delta}{4}})$, vanishing at the origin and satisfying the estimate*

$$|\psi(z_1, z_2)| \leq \delta^{-\nu} \max_{(t_1,t_2)\in\overline{\prod}(\rho)} |\gamma(t_1, t_2)|, \qquad (z_1, z_2) \in \overline{\Omega}(\rho e^{-\frac{\delta}{4}}).$$

Proof. Let

$$P_s = \{ z \in \mathbb{C}^1 : |z - 2\pi s| = 10^{-1} \}, \quad s = 0, \pm 1, \ldots$$

and let

$$C_0 = \{ z \in \mathbb{C}^1 : |z| = 1 \},\ C_k = \{ z \in C^1 : |z| = k e^{\frac{\delta}{4}} \rho^{-1} \}, \quad k = 1, 2, \ldots$$

In the case $C_k \bigcap P_s = \emptyset$ for all $s \in \mathbb{Z}$ we set $L_k = C_k$. Otherwise there may be exactly two exceptional circles $P_{\pm s}$ intersecting C_k and we denote by L_k the closed contour composed by a part of C_k outside the set $\bigcup_s P_s$ and by parts of $P_{\pm s}$ located outside C_k. It is evident that every contour L_k is symmetric with respect to the real line and its length does not exceed $2\pi(k e^{\frac{\delta}{4}} \rho^{-1} + 1)$.

With the system of contours $\{L_k\}_0^\infty$ being chosen, let us set, following [43],

$$\omega(z_1, z_2) = \sum_{k=0}^{\infty} \frac{1}{2\pi i} \frac{\partial^k \gamma(0, z_2)}{\partial z_1^k} \oint_{L_k} \frac{e^{\lambda z_1} d\lambda}{\lambda^{k+1}(e^\lambda - 1)} \tag{3.1.9}$$

and show that ω is analytic in the domain $\Omega(\rho e^{-\frac{\delta}{4}})$. Indeed, for $\lambda \in \mathbb{C}$ the estimates

$$|e^{\lambda z_1}| \leq e^{(H(\lambda)+\chi)|\lambda|}, \quad \chi = \rho e^{-\frac{\delta}{4}},\ z_1 = t + \xi,\ |\xi| < \chi,$$

and

$$|e^\lambda - 1|^{-1} \leq q e^{-H(\lambda)|\lambda|}, \quad \lambda \in L_k,$$

hold with a constant $q > 0$ and

$$H(\lambda) = \begin{cases} \cos(\arg\lambda), & \mathrm{Re}\lambda \geq 0, \\ 0, & \mathrm{Re}\lambda \leq 0. \end{cases}$$

If $M = \max_{z \in \overline{\prod}(\rho)} |\gamma(z)|$, then according to the Cauchy inequalities we have

$$\left| \frac{\partial^k \gamma(0, z_2)}{\partial z_1^k} \right| \leq \frac{k!}{\rho^k} M, \quad |z_2| \leq \rho.$$

If $\lambda \in L_k$, then $|\lambda| \leq k e^{\frac{\delta}{4}} \rho^{-1} + 1$. Hence for $(z_1, z_2) \in \overline{\Omega}(\rho e^{-\frac{\delta}{4}})$ we have

$$\left| \frac{1}{2\pi i} \frac{\partial^k \gamma(0, z_2)}{\partial z_1^k} \oint_{L_k} \frac{e^{\lambda z_1} d\lambda}{\lambda^{k+1}(e^\lambda - 1)} \right| \leq M e^2 q \frac{k!}{\rho^k} \frac{e^{\chi R_k}}{R_k^k} \leq C k e^{-\frac{k\delta}{4}} M,$$

where $R_k = k e^{\frac{\delta}{4}} \rho^{-1}$, $k \geq 1$, and

$$C = e^2 q \sup_{k \geq 1} k! k^{-(k+\frac{1}{2})} e^k.$$

Therefore the series (3.1.9) uniformly converges for $(z_1, z_2) \in \overline{\Omega}(\rho e^{-\frac{\delta}{4}})$ and defines a function $\omega(z_1, z_2)$ continuous in $\overline{\Omega}(\rho e^{-\frac{\delta}{4}})$, analytic in $\Omega(\rho e^{-\frac{\delta}{4}})$ and satisfying the estimate

$$|\omega(z_1, z_2)| \leq C(1 + \sum_{k=1}^{\infty} k e^{-k\frac{\delta}{4}}) M \leq C(1 + (1 - e^{-\frac{\delta}{4}})^{-2}) M, \quad (z_1, z_2) \in \overline{\Omega}(\rho e^{-\frac{\delta}{4}}).$$

For $(z_1, z_2) \in \prod(\rho e^{-\frac{\delta}{4}})$ we have

$$\omega(z_1 + 1, z_2) = \sum_{k=0}^{\infty} \frac{1}{2\pi i} \frac{\partial^k \gamma(0, z_2)}{\partial z_1^k} \oint_{L_k} \frac{e^{\lambda z_1} e^\lambda d\lambda}{\lambda^{k+1}(e^\lambda - 1)} d\lambda$$

$$= \sum_{k=0}^{\infty} \frac{1}{2\pi i} \frac{\partial^k \gamma(0, z_2)}{\partial z_1^k} \oint_{L_k} \frac{e^{\lambda z_1}(e^\lambda - 1)}{\lambda^{k+1}(e^\lambda - 1)} d\lambda$$

$$+ \sum_{k=0}^{\infty} \frac{1}{2\pi i} \frac{\partial^k \gamma(0, z_2)}{\partial z_1^k} \oint_{L_k} \frac{e^{\lambda z_1} d\lambda}{\lambda^{k+1}(e^\lambda - 1)}$$

$$= \sum_{k=0}^{\infty} \frac{1}{2\pi i} \frac{\partial^k \gamma(0, z_2)}{\partial z_1^k} \frac{2\pi i z_1^k}{k!} + \omega(z_1, z_2) = \gamma(z_1, z_2) + \omega(z_1, z_2),$$

which means that $\omega(z_1, z_2)$ is a solution of (3.1.8).

The function

$$\psi(z_1, z_2) = \omega(z_1, z_2) - \omega(0,0)$$

satisfies (3.1.8) as well and $\psi(0,0) = 0$. At last, we can choose a number $\nu > 0$ sufficiently large to satisfy the inequality

$$2C(1 + (1 - e^{-\frac{\delta}{4}})^{-2}) < \delta^{-\nu}$$

for all $\delta \in (0, 2^{-1})$.

Thus the function ψ has all properties from the statement of Lemma 3.2. □

Let us note in addition that $\psi(z_1, z_2)$ is real for real z_1, z_2.

Now we construct appropriate solutions of (3.1.4) and (3.1.5).

Lemma 3.3. *There exist numbers ν and $\alpha_0 > 4$ such that if the functions $f(z)$ and $g(z_1, z_2)$ are analytic in $\Delta(\rho)$ and $\prod(\rho)$, $\rho \le 2^{-1}$, respectively, and satisfy the estimates*

$$\max_{z\in\overline{\Delta}(\rho)} |f(z)| \le \rho\delta^{\alpha}, \qquad \max_{(z_1,z_2)\in\overline{\prod}(\rho)} |g(z_1, z_2)| \le \rho\delta^{\alpha}, \quad 0 < \delta \le 2^{-2}, \ \alpha > \alpha_0,$$

then there exist solutions $\varphi(z)$ and $t(z_1, z_2)$ of equations (3.1.4) and (3.1.5) analytic in the domains $I(\rho e^{-\frac{\delta}{4}})$ and $\Omega(\rho e^{-\frac{\delta}{4}})$, respectively, and satisfying conditions $\varphi(0) = 0$, $t(0,0) = 0$, and the estimates

$$\max_{z\in\overline{I}(\rho e^{-\frac{\delta}{4}})} |\varphi(z)| \le \delta^{-\nu} \max_{z\in\overline{\Delta}(\rho)} |f(z)|,$$

$$\max_{(z_1,z_2)\in\overline{\Omega}(\rho e^{-\frac{\delta}{4}})} |t(z_1, z_2)| \le \delta^{-\nu} \max_{(z_1,z_2)\in\overline{\prod}(\rho)} |g(z_1, z_2)|,$$

hold. The functions $f_1(z)$ and $g_1(z_1, z_2)$ defined by (3.1.6) and (3.1.7) satisfy the estimates

$$\max_{z\in\overline{\Delta}(\rho e^{-\delta})} |f_1(z)| \le \rho^{-1}\delta^{-\alpha} \max_{z\in\overline{\Delta}(\rho)} |f(z)|^2,$$

$$\max_{(z_1,z_2)\in\overline{\prod}(\rho e^{-\delta})} |g_1(z_1, z_2)| \le \rho^{-1}\delta^{-\alpha} \max_{(z_1,z_2)\in\overline{\prod}(\rho)} |g(z_1, z_2)|^2.$$

Proof. A number ν and solutions $\varphi(z)$ and $t(z_1, z_2)$ with the above properties exist according to Lemma 3.2.

To prove the estimates for $f_1(z)$ and $g_1(z_1, z_2)$ we note that according to the Cauchy inequalities the functions $\varphi(z)$ and $t(z_1, z_2)$ satisfy the estimates

$$|\varphi'(z)| \le C\rho^{-1}\delta^{-\nu-2} \max_{t\in\overline{\Delta}(\rho)} |f(t)|, \quad z \in \overline{I}(\rho e^{-\frac{\delta}{2}}),$$

and

$$\left|\frac{\partial t(z_1,z_2)}{\partial z_1}\right| + \left|\frac{\partial t(z_1,z_2)}{\partial z_2}\right|$$

$$\leq C\rho^{-1}\delta^{-\nu-2} \max_{z\in\overline{\prod}(\rho)} |g(z_1,z_2)|, \quad (z_1,z_2) \in \overline{\Omega}(\rho e^{-\frac{\delta}{2}}),$$

with

$$C = 2 \max_{0\leq\delta\leq 1} \delta^2(1 - e^{-\frac{\delta}{4}})^{-2} > 2.$$

Let us now fix $\alpha_0 = \nu + 4 + \log_2 C > 4$. Then for every $\delta \in (0, 2^{-1}]$ and every $\alpha \geq \alpha_0$ we have

$$C\delta^{-\nu-2} \leq C\delta^{-\alpha}2^{-\alpha_0+\nu+2} \leq \delta^{-\alpha}. \tag{3.1.10}$$

An elementary calculation shows that for $\delta \in (0, 2^{-2}]$ the inequalities

$$\delta^4 + e^{-\frac{3}{4}\delta} \leq e^{-\frac{1}{2}\delta}, \; \delta^4 + e^{-\delta} \leq e^{-\frac{3}{4}\delta}$$

hold, and if we assume

$$|f(z)| \leq \rho\delta^{\alpha}, \; z \in \overline{\Delta}(\rho); \quad |g(z_1,z_2)| \leq \rho\delta^{\alpha}, \; (z_1,z_2) \in \overline{\prod}(\rho),$$

then we obtain

$$z + 1 + f(z) \in \overline{I}(\rho e^{-\frac{\delta}{2}}), \qquad z \in \overline{\Delta}(\rho e^{-\frac{3}{4}\delta});$$

$$z_2 + g(z_1,z_2)) \in \overline{\Omega}(\rho e^{-\frac{1}{2}\delta}), \qquad (z_1,z_2) \in \overline{\prod}(\rho e^{-\frac{3}{4}\delta}).$$

Therefore

$$|\varphi(z + 1 + f(z)) - \varphi(z + 1)|$$

$$\leq C\rho^{-1}\delta^{-\nu-2} \max_{z\in\overline{\Delta}(\rho)} |f(t)|^2, \; z \in \overline{\Delta}(\rho e^{-\frac{3}{4}\delta}),$$

and

$$|t(z_1 + 1 + f(z_1), z_2 + g(z_1,z_2)) - t(z_1 + 1, z_2)|$$

$$\leq C\rho^{-1}\delta^{-\nu-2} \max_{(t_1,t_2)\in\overline{\prod}(\rho)} |g(t_1,t_2)|^2, \; (z_1,z_2) \in \overline{\prod}(\rho e^{-\frac{3}{4}\delta}),$$

implying

$$|f_1(z + \varphi(z))| \leq C\rho^{-1}\delta^{-\nu-2} \max_{t\in\overline{\Delta}(\rho)} |f(t)|^2, \; z \in \overline{\Delta}(\rho e^{-\frac{3}{4}\delta}),$$

and

$$|(g_1(z_1 + \varphi(z_1), z_2 + t(z_1,z_2))|$$

$$\leq C\rho^{-1}\delta^{-\nu-2} \max_{(t_1,t_2)\in\overline{\prod}(\rho)} |g(t_1,t_2)|^2, \quad (z_1,z_2) \in \overline{\prod}(\rho e^{-\frac{3}{4}\delta}).$$

For every $\alpha > \alpha_0$ we have $|\varphi(z)| \le \rho\delta^{\alpha-\nu} \le \rho\delta^4$, $z \in \overline{\Delta}(\rho e^{-\frac{1}{4}\delta})$. If $|t| = \rho e^{-\frac{3}{4}\delta}$, then

$$|t + \varphi(t)| \ge \rho e^{-\frac{3}{4}\delta} - \max_{t \in \overline{\Delta}(\rho e^{-\frac{1}{4}\delta})} |\varphi(t)| \ge \rho(e^{-\frac{3}{4}\delta} - \delta^4) \ge \rho e^{-\delta},$$

and the image of the disc $\overline{\Delta}(\rho e^{-\frac{3}{4}\delta})$ with respect to the mapping $z + \varphi(z)$ contains the disc $\overline{\Delta}(\rho e^{-\delta})$. We conclude that for every $z \in \overline{\Delta}(\rho e^{-\delta})$ there exists the unique solution $t = t(z)$ of the equation

$$z = t + \varphi(t), \quad t \in \overline{\Delta}(\rho e^{-\frac{3}{4}\delta}).$$

Hence

$$|f_1(z)| \le \max_{t \in \Omega(\rho e^{-\frac{3}{4}\delta})} |f_1(t + \varphi(t))| \le C\rho^{-1}\delta^{-\nu-2} \max_{t \in \overline{\Delta}(\rho)} |f(t)|^2, \ z \in \overline{\Delta}(\rho e^{-\delta}).$$

Similarly, the estimate

$$|g_1(z_1, z_2)| \le C\rho^{-1}\delta^{-\nu-2} \max_{(t_1,t_2) \in \overline{\prod}(\rho)} |g(t_1, t_2)|^2, \quad (z_1, z_2) \in \overline{\Pi}(\rho e^{-\delta}),$$

holds and using (3.1.10) we complete the proof of Lemma 3.3. □

To continue the proof of Theorem 3.1 in the analytic case we set $N = 2\alpha$ where $\alpha > 4$ is a number from Lemma 3.3 and fix a small number $\delta_0 < 2^{-3}$ such that the sequence

$$\delta_0 < (\tfrac{1}{2})^3, \ \delta_1 = \delta_0^{\frac{3}{2}}, \ \dots, \ \delta_s = \delta_{s-1}^{\frac{3}{2}}, \dots$$

satisfies the inequality

$$\sum_{s=0}^{\infty} \delta_s < 2 \ln 2.$$

Since $f(0) = f'(0) = 0$, $g(0,0) = g'_x(0,0) = g'_y(0,0) = 0$, then

$$\max_{z \in \overline{\Delta}(\rho)} |f(z)| \le K\rho^2, \qquad \max_{(z_1,z_2) \in \overline{\prod}(\rho)} |g(z_1, z_2)| \le K\rho^2$$

with a number K not depending on ρ. We choose ρ_0 such that $K\rho_0 \le \delta_0^N$ and define the sequences

$$\rho_0 < (\tfrac{1}{2})^3, \ \rho_1 = \rho_0 e^{-\delta_0}, \ \dots, \ \rho_s = \rho_{s-1} e^{-\delta_{s-1}}, \dots,$$

$$M_0 = \delta_0^N, \ M_1 = M_0^{\frac{3}{2}}, \ \dots, M_s = M_{s-1}^{\frac{3}{2}} = \delta_s^N \dots.$$

It is evident that $\rho_s \ge 2^{-1}\rho_0, s = 1, 2, \dots.$

Now we set $f_0(z) = f(z)$, $z \in \overline{\Delta}(\rho_0)$, and $g_0(z_1, z_2) = g(z_1, z_2)$, $(z_1, z_2) \in \overline{\prod}(\rho_0)$, and obtain

$$\max_{z\in\overline{\Delta}(\rho_0)} |f_0(z)| \le \rho_0\delta_0^N, \qquad \max_{(z_1,z_2)\in\overline{\prod}(\rho_0)} |g_0(z_1, z_2)| \le \rho_0\delta_0^N.$$

Assume that we have already found functions $f_s(z)$, $z \in \overline{\Delta}(\rho_s)$, and $g_s(z_1, z_2)$, $z \in \overline{\prod}(\rho_s)$, satisfying

$$|f_s(z)| \le \rho_s\delta_s^N, \quad z \in \overline{\Delta}(\rho_s); \quad |g_s(z_1, z_2)| \le \rho_s\delta_s^N, \quad z \in \overline{\Delta}(\rho_s).$$

Then we apply Lemma 3.3 with $f = f_s$, $g = g_s$, $\rho = \rho_s$, $\delta = \delta_s$, $\alpha = N$, find solutions φ_s, t_s of (3.1.4) and (3.1.5) and functions f_{s+1}, g_{s+1} defined by (3.1.6) and (3.1.7), respectively. According to Lemma 3.3 we have

$$|\varphi_s(z)| \le \rho_s\delta_s^{N-\nu}, \ z \in \overline{I}(\rho_{s+1}),$$

$$|t_s(z_1, z_2)| \le \rho_s\delta_s^{N-\nu}, z \in \overline{\Omega}(\rho_{s+1}), \ s = 0, 1, \ldots$$

and

$$|f_{s+1}(z)| \le \rho_s^{-1}\delta_s^{-\alpha}\rho_s^2\delta_s^{2N} = \rho_s\delta_s^{2N-\alpha} \le \rho_{s+1}\delta_{s+1}^N, \quad z \in \overline{\Delta}(\rho_{s+1}),$$

$$|g_{s+1}(z_1, z_2)| \le \rho_{s+1}\delta_{s+1}^N, \qquad (z_1, z_2) \in \overline{\prod}(\rho_{s+1}).$$

Furthermore, let us define a sequence of mappings

$$\Phi_s(x, y) = (x + \varphi_s(x), \, y + t_s(x, y)), \quad s = 0, 1, \ldots,$$

and set

$$\Psi_s = \Phi_s\Phi_{s-1} \ldots \Phi_0.$$

If $\Psi_s(x, y) = (H_s(x), \, Q_s(x, y))$, then

$$H_{s+1}(x) = H_s(x) + \varphi_{s+1}(H_s(x))$$

and

$$Q_{s+1}(x, y) = Q_s(x, y) + t_{s+1}(H_s(x), \, Q_s(x, y)).$$

Hence

$$H_{s+1}(x) - H_s(x) = \varphi_{s+1}(H_s(x)), \qquad s = 0, 1, \ldots,$$

$$Q_{s+1}(x, y) - Q_s(x, y) = t_{s+1}(H_s(x), \, Q_s(x, y)), \qquad s = 0, 1, \ldots.$$

Let us prove the estimates

$$|H_s(z) - z| \le \frac{\rho_0}{2}, \quad z \in \overline{I}\left(\frac{\rho_0}{2}\right), \tag{3.1.11}$$

for $s = 0, 1, \ldots$. For $s = 0$ we have

$$|H_0(z) - z| = |\varphi_0(z)| \le \rho_0 \delta_0^{N-\nu} \le \rho_0 \delta_0^4,$$

since $N - \nu > 4$. Suppose that the estimate

$$|H_s(z) - z| \le \sum_{k=0}^{s} \rho_k \delta_k^4, \quad z \in \overline{I}\left(\frac{\rho_0}{2}\right), \tag{3.1.12}$$

has already been proved. Then

$$|H_{s+1}(z) - z| = |H_{s+1}(z) - H_s(z)| + |H_s(z) - z|$$

$$\le |\varphi_{s+1}(H_s(z))| + \sum_{k=0}^{s} \rho_k \delta_k^4, \quad z \in \overline{I}\left(\frac{\rho_0}{2}\right).$$

It is easy to see that since $\delta_0 < 2^{-3}$, $\rho_0 < 2^{-3}$, we have

$$\sum_{k=0}^{\infty} \rho_k \delta_k^4 < \frac{\rho_0}{4}, \quad \sum_{k=0}^{\infty} \delta_k < \frac{1}{4},$$

and hence

$$\rho_s = \rho_0 e^{-\sum_{k=0}^{s-1} \delta_k} \ge \rho_0 e^{-\frac{1}{4}} > \frac{3\rho_0}{4}.$$

Now (3.1.12) implies

$$|H_s(z)| \le \frac{\rho_0}{2} + \sum_{k=0}^{s} \rho_k \delta_k^4 \le \frac{3\rho_0}{4} < \rho_{s+2}, \quad z \in \overline{I}\left(\frac{\rho_0}{2}\right).$$

Therefore H_s maps $I\left(2^{-1}\rho_0\right)$ into $I(\rho_{s+2})$, $|\varphi_{s+1}(H_s(z))| \le \rho_{s+1}\delta_{s+1}^4$ and

$$|H_{s+1}(z) - z| \le \sum_{k=0}^{s+1} \rho_k \delta_k^4, \qquad z \in \overline{I}\left(\frac{\rho_0}{2}\right),$$

proving (3.1.11) for all $s \ge 0$. Since

$$|H_{s+1}(z) - H_s(z)| = |\varphi_{s+1}(H_s(z))| \le \rho_{s+1}\delta_{s+1}^{N-\nu}, \quad z \in \overline{I}\left(\frac{\rho_0}{2}\right),$$

the sequence H_s converges uniformly in $\overline{I}(2^{-1}\rho_0)$ and $H_\infty = \lim_{s\to\infty} H_s$ is a function analytic in $I(2^{-1}\rho_0)$ and continuous in $\overline{I}(2^{-1}\rho_0)$. Besides,

$$|H_\infty(z) - z| \le \sum_{k=0}^{\infty} \rho_k \delta_k^4 \le \rho_0 \sum_{k=0}^{\infty} \delta_k, \quad z \in \overline{I}\left(\frac{\rho_0}{2}\right).$$

The Cauchy inequality yields

$$|H'_\infty(z) - 1| \le 4\sum_{k=0}^{\infty} \delta_k < 1, \quad z \in \overline{I}\left(\frac{\rho_0}{8}\right),$$

and we find that H_∞ is an analytic diffeomorphism of a neighborhood of $[0,1] \subset \mathbb{R}^2$ onto its image.

In the same way we prove the estimates

$$|Q_s(z_1,z_2) - z_2| \le \sum_{k=0}^{s} \rho_k \delta_k^4, \quad (z_1,z_2) \in \overline{\Omega}\left(\frac{\rho_0}{2}\right),$$

$$|Q_{s+1}(z_1,z_2) - Q_s(z_1,z_2)| \le \rho_{s+1}\delta_{s+1}^{N-\nu}, \quad (z_1,z_2) \in \overline{\Omega}\left(\frac{\rho_0}{2}\right),$$

and conclude that sequence $\{Q_s(z_1,z_2)\}_0^\infty$ converges to a function $Q_\infty(z_1,z_2)$ analytic in $\overline{\Omega}(2^{-1}\rho_0)$. The estimates

$$|Q_\infty(z_1,z_2) - z_2| \le \rho_0\sum_{k=0}^{\infty}\delta_k, \quad \left|\frac{\partial Q_\infty(z_1,z_2)}{\partial z_1} - 1\right| \le 4\sum_{k=0}^{\infty}\delta_k < 1,$$

show that the limit mapping $\Psi_\infty(z_1,z_2) = (H_\infty(z_1), Q_\infty(z_1,z_2))$ is invertible in $\Omega(8^{-1}\rho_0)$ and the equalities

$$\Psi_s G \Psi_s^{-1}(z_1,z_2) = G_s(z_1,z_2), \quad (z_1,z_2) \in \overline{\Omega}\left(\frac{\rho_0}{8}\right), \quad s \ge 0,$$

yield

$$\Psi_\infty G \Psi_\infty^{-1}(z_1,z_2) = (z_1 + 1, z_2).$$

This completes the proof of Theorem 3.1 in the analytic case. □

3.1.3 Local solvability at an isolated fixed point

If x_0 is an isolated fixed point for F and equation (3.0.1) has a local solution $\varphi(x)$ at x_0, then the equation

$$y = g(x_0, y) \tag{3.1.13}$$

has the solution $y = \varphi(x_0)$. The following example shows that even if a solution of (3.1.13) exists and the mapping $G(x,y) = (F(x), g(x,y))$ is a C^k-diffeomorphism with an isolated fixed point (x_0,y_0) it may happen that (3.0.1) has no local solutions at x_0.

Example 3.3. Consider the equation

$$\varphi(x + x^3) - \varphi(x) = x^2$$

with the analytic diffeomorphism $F(x) = x + x^3$. Here $g(x,y) = y + x^2$ and $y_0 = 0$ satisfies (3.1.13) at the fixed point $x_0 = 0$. However equation (3.1.13) has no local continuous solutions. Indeed, the representation

$$F^{-1}(x) = x - x^3 + O(x^5) = x(1 + 2x^2)^{-\frac{1}{2}} + O(x^4)$$

shows that the fixed point $x_0 = 0$ is attracting for F^{-1} and according to (2.3.3) we have

$$\lim_{n\to\infty} \sqrt{n}F^{-n}(x) = \frac{\sqrt{2}}{2}, \quad x > 0.$$

If φ is a local solution, then

$$\varphi(x) = \varphi(F^{-n}(x)) + \sum_{k=1}^{n}(F^{-k}(x))^2, \quad n = 1, 2, \ldots ,$$

and since $\sum_{k=1}^{\infty}(F^{-k}(x))^2 = \infty$, the function φ cannot be bounded at $x = 0$.

If $F(x_0) \neq x_0$, then Corollary 3.1 describes all solutions of equation (3.0.1) in terms of a transformation which reduces it to the simplest form $\varphi(x+1) = \pm\varphi(x)$. The problem of a similar description of solutions to the same equation in a neighborhood of a fixed point is essentially more complicated. In particular, it contains the problem of a normal form of a local two-dimensional mapping at a fixed point. Here we restrict ourselves to a statement of solvability conditions only. In Section 3.2 below we describe the set of solutions in the presence of fixed points for some linear equations.

To formulate solvability conditions at a fixed point, let us first consider the problem of a formal solvability. Assume that $y_0 = g(x_0, y_0)$. If φ is a local C^k-solution of (3.0.1), $\varphi(x_0) = y_0$, then

$$\frac{d^j}{dx^j}(\varphi(F(x)) - g(x, \varphi(x))\,|_{x=x_0} = 0, \quad j = 1, 2, \ldots, k. \tag{3.1.14}$$

If we substitute $\varphi^{(j)}(x_0) = c_j$, we will arrive at a system of algebraic equations with respect to $c_0, c_1, \ldots$, which is finite if $k < \infty$ and infinite otherwise. Write (3.1.14) in the recurrent form

$$c_0 = y_0, \quad c_j(\lambda^j - \mu) = R_j(c_0, c_1, \ldots, c_{j-1}), \quad j = 1, 2, \ldots, k, \tag{3.1.15}$$

with some polynomials R_j and

$$\lambda = F'(x_0), \quad \mu = \frac{\partial g(x_0, y_0)}{\partial y}, \quad R_1 = \frac{\partial g(x_0, y_0)}{\partial x}. \tag{3.1.16}$$

If

$$\lambda^j \neq \mu, \quad j = 1, \ldots k, \tag{3.1.17}$$

then the numbers

$$c_j = \frac{1}{\lambda^j - \mu} R_j(c_1, \ldots, c_{j-1}),$$

define the unique solution of (3.1.15). In the general case, we say that (3.0.1) is *formally* C^k*-solvable* (just *formally solvable* if $k = \infty$ or $k = \aleph$) at x_0 if the system (3.1.15) has a solution $c_1, \ldots, c_k$. The corresponding Taylor polynomial

$$\varphi_0(x) = y_0 + \sum_{j=1}^{k} \frac{1}{j!} c_j (x - x_0)^j$$

is called a *formal* C^k*-solution* at x_0. In the C^∞ — or analytic — case the series

$$\hat{\varphi} = y_0 + \sum_{j=0}^{\infty} \frac{1}{j!} c_j (x - x_0)^j$$

is called a *formal solution.* The *non-resonance conditions* (3.1.17) are sufficient for (3.0.1) to have a formal solution at x_0.

The following proposition states some conditions of the solvability of equation (3.0.1) for functions of a finite smoothness.

Theorem 3.2. *Let the mapping* $G(x,y) = (F(x), g(x,y))$ *with a fixed point* (x_0, y_0) *belong to a class* C^k, $1 \leq k < \infty$, *and let all derivatives of the function* $g(x,y)$ *of order* k *satisfy the Lipschitz condition*

$$\left| \frac{\partial^k g(x, y_1)}{\partial x^m \partial y^{k-m}} - \frac{\partial^k g(x, y_2)}{\partial x^m \partial y^{k-m}} \right| \leq c|y_1 - y_2|, \quad m = 0, \ldots, k.$$

If

$$|F'(x_0)|^k \neq \left| \frac{\partial g(x_0, y_0)}{\partial y} \right|, \tag{3.1.18}$$

then for every formal C^k*-solution* φ_0 *of* (3.0.1) *at* x_0 *there exists a local* C^k*-solution* φ *whose Taylor polynomial of order k at* x_0 *coincides with* φ_0.

As Example 3.3 shows, if (3.1.18) is not fulfilled, then the formal C^k-solvability does not imply the C^k-solvability: here $k = 1$, the zero polynomial is a formal C^1-solution, but no continuous solution exists.

Proof of Theorem 3.2. We essentially use arguments from the proof of Theorem 2.5. Without loss of generality we assume $x_0 = 0$.

If φ_0 is a formal C^k-solution at $x_0 = 0$, then a C^k-solution φ must be of the form $\varphi(x) = \varphi_0(x) + \psi(x)$ where ψ is a C^k-function with the zero Taylor polynomial of order k, which satisfies the equation

$$\psi(F(x)) = \tilde{g}(x, \psi(x)) + \gamma(x)$$

where

$$\tilde{g}(x,y) = g(x,y + \varphi_0(x)) - g(x,\varphi_0(x)),$$

the function $\gamma(x) = g(x,\varphi_0(x)) - \varphi_0(F(x))$ belongs to C^k and all its derivatives up to the order k vanish at the origin.

Let λ and μ be defined by (3.1.16). In the case $|\lambda|^k < |\mu|$ we have $\mu = \tilde{g}'_y(0,0) \neq 0$, and by the Implicit Function Theorem there exists a C^k-function $h(x,z)$, $h(0,0) = 0$, such that

$$\tilde{g}(x,h(x,z)) = z, \quad h(x,\tilde{g}(x,y)) = y.$$

We set $\psi(x) = h(x,z(x))$, and arrive at the equation

$$z(x) = h(F(x),z(F(x))) - \gamma(x)$$

with $h'_y(0,0) = \mu^{-1}$ and $|\mu^{-1}||\lambda|^k < 1$.

In the case $|\lambda|^k > |\mu|$ the mapping F is a local C^k-diffeomorphism and $z(x) = \psi(F(x))$ satisfies the equation

$$z(x) = \tilde{g}(x,z(F^{-1}(x))) + \gamma(x)$$

with $|\mu||\lambda|^{-k} < 1$.

The previous arguments show that to prove Theorem 3.2 it is sufficient to prove the local C^k-solvability of the equation

$$\psi(x) = g(x,\psi(F(x)) + \gamma(x) \tag{3.1.19}$$

with

$$g(x,0) \equiv 0, \quad \gamma^{(j)}(0) = 0,\ j = 0,1,\ \dots\ ,k,$$

and $|\mu||\lambda|^k < 1$ where, as above,

$$\mu = \frac{\partial g(0,0)}{\partial y}, \quad \lambda = F'(0).$$

Let us first assume $|\lambda| < 1$. In this case there exists $\varepsilon_0 > 0$ such that for every $\epsilon < \epsilon_0$ the interval $I(\epsilon) = [-\varepsilon,\varepsilon]$ is F-invariant and therefore the operator

$$T\psi(x) = g(x,\psi(F(x))) + \gamma(x)$$

is well defined in the space $\mathcal{F}(\epsilon)$ of all C^k-functions on $I(\epsilon)$ satisfying $\psi(0) = \cdots = \psi^{(k)}(0) = 0$. Being endowed with the norm

$$\|\psi\| = \max_{x\in I(\epsilon)} |\psi^{(k)}(x)|,$$

$\mathcal{F}(\epsilon)$ becomes a Banach space. It is evident that the function γ belongs to the space $\mathcal{F}(\epsilon)$ for every $\epsilon < \epsilon_0$ and its norm does not exceed

$$\|\gamma\|_0 = \max_{x\in I(\epsilon_0)} |\gamma^{(k)}(x)|.$$

Given $M > 0$, we introduce the ball

$$B(M,\epsilon) = \{\psi \in \mathcal{F}(\epsilon) : \|\psi\| \le M\}.$$

Our aim is to find M and ϵ such that T is a contractive operator in $B(M,\epsilon)$.

For every $k \ge 1$ we have

$$(T\psi)^{(k)}(x) = \frac{\partial g(x,\psi(F(x)))}{\partial y}\psi^{(k)}(F(x))(F'(x))^k$$

$$+ R_k(x,\psi(F(x)), \ldots, \psi^{(k-1)}(F(x))) + \gamma^{(k)}(x).$$

Here $R_k(x,y_0, \ldots, y_{k-1})$ is a continuous function polynomially depending on $y_1, \ldots, y_{k-1}$. Besides, $R_k(x,0,\ldots,0) = 0$ and the Lipschitz condition is fulfilled: for every $N > 0$ there exists a constant $c = c(N)$ such that $|u_j| \le N$, $|v_j| \le N$ imply

$$|R_k(x,u_0, \ldots, u_{k-1}) - R_k(x,v_0, \ldots, v_{k-1})|$$
$$\le c(N) \max_{0\le j\le k-1} |u_j - v_j|. \tag{3.1.20}$$

Since

$$q = \left|\frac{\partial g(0,0)}{\partial y}\right| |F'(0)|^k < 1,$$

there exists $\sigma > 0$ such that $q + 3\sigma < 1$. We fix $M \ge \sigma^{-1}\|\gamma\|_0$ and choose $\epsilon > 0$ and $\delta > 0$ to satisfy the inequalities

$$\left|\frac{\partial g(x,y)}{\partial y}\right| |F'(x)|^k \le q < 1, \quad |x| \le \varepsilon, \ |y| \le \delta,$$

and

$$\epsilon^k M \le \delta, \qquad \epsilon c(M) \le \sigma$$

where $c(M)$ is the constant from (3.1.20) with $N = M$. If now $\psi \in B(M,\epsilon)$, then

$$|\psi^{(j)}(x)| \le \varepsilon^{k-j}M, \quad j = 0,1, \ldots, k,$$

and

$$|(T\psi)^{(k)}(x)| \le (q+\sigma)M + \varepsilon c(M)M + \|\gamma\| \le (q+3\sigma)M.$$

Therefore the ball $B(M,\epsilon)$ is T-invariant.

Furthermore, if $\psi_1, \psi_2 \in B(M)$, then the estimates (3.1.20) and

$$|\psi_1^{(j)}(x) - \psi_2^{(j)}(x)| \le \varepsilon^{k-j}\|\psi_1 - \psi_2\|, \qquad j = 0,1,\ldots,k,$$

yield

$$\|T\psi_1 - T\psi_2\| \le (q+\sigma)\|\psi_1 - \psi_2\| + c(M)\varepsilon\|\psi_1 - \psi_2\|$$

$$\le (q+2\sigma)\|\psi_1 - \psi_2\|.$$

Hence T contracts the ball $B(M,\epsilon)$ and equation (3.1.19) has a solution $\psi \in B(M,\epsilon)$.

Now consider the case $|\lambda| \geq 1$. Generally speaking, no interval $I(\epsilon)$ is invariant with respect to F and we need an alternative construction.

Note that the inequality $|\mu||\lambda|^k < 1$ implies $|\mu||\lambda|^i < 1$, $i = 0, \ldots, k$. Hence there exist numbers $q \in (0,1)$, $\varepsilon > 0$, $\delta > 0$ such that

$$\left|\frac{\partial g(x,y)}{\partial y}\right| |F'(x)|^i \leq q < 1, \quad i = 0, \ldots, k, \; |x| < \varepsilon, \; |y| < \delta,$$

and

$$|\gamma(x)| \leq (1-q)\delta, \qquad |x| \leq \epsilon.$$

With $\varepsilon > 0, \delta > 0$ being fixed we choose a C^∞-function $\tau : \mathbb{R} \to [0,1]$ equal to 1 in a neighborhood of the origin and vanishing as $|x| \geq \varepsilon$. As above, let $\mathcal{F}(\epsilon)$ be the space of all C^k-functions defined on $I(\epsilon)$ with $\psi(0) = \cdots = \psi^{(k)}(0) = 0$. Then the operator

$$T\psi(x) = \begin{cases} \tau(F(x))(g(x,\psi(F(x))) + \gamma(x)), & |F(x)| \leq \varepsilon, \\ 0, & |F(x)| \geq \varepsilon, \end{cases}$$

is defined in $\mathcal{F}(\epsilon)$. It is sufficient to prove the existence of a solution $\psi \in \mathcal{F}(\epsilon)$ for the equation

$$\psi = T\psi.$$

Denote by $B(M_1, \ldots, M_k)$ the set of all functions $\psi \in \mathcal{F}(\epsilon)$ such that

$$|\psi(x)| \leq \delta, \; |\psi^{(i)}(x)| \leq M_i, \quad i = 1, \ldots, k.$$

If $\psi \in B(M_1, \ldots, M_k)$, then

$$|T\psi(x)| \leq |g(x,\psi(F(x)))| + |\gamma(x)| \leq q\delta + |\gamma(x)| \leq \delta$$

and

$$|(T\psi)^{(i)}(x)| \leq qM_i + R_i(M_1, \ldots, M_{i-1}), \quad i = 1, \ldots, k,$$

with some polynomials $R_i, \deg R_i \leq (i-1)$. We choose recurrently numbers M_i such that

$$M_i > \frac{R_i(M_1, \ldots, M_{i-1})}{1-q}, \quad i = 1, \ldots, k,$$

and find that $B(M_1, \ldots, M_k)$ is T-invariant.

For $\psi_1, \psi_2 \in B(M_1, \ldots, M_k)$, we have

$$|T\psi_1(x) - T\psi_2(x)| \leq qd_0$$

and

$$|(T\psi_1)^{(i)}(x) - (T\psi_2)^{(i)}(x)| \leq qd_i + c\sum_{j=0}^{i-1} d_j, \quad i = 1, \ldots, k,$$

where

$$d_j = \max_x |\psi_1^{(j)}(x) - \psi_2^{(j)}(x)|$$

and c is a constant depending on $M_1, \ldots, M_k$ only. Now we choose numbers $\alpha_k = 1, \alpha_j > 0$, such that

$$\tilde{q} = q + \frac{c}{\alpha_i} \sum_{j=i+1}^{k} \alpha_j < 1, \quad i = 1, 2, \ldots, k-1,$$

and define the metric in $B(M_1, \ldots, M_k)$ by the formula

$$\operatorname{dist}(\psi_1, \psi_2) = \sum_{i=0}^{k} \alpha_i d_i.$$

With the above choice of $\alpha_0, \ldots, \alpha_k$ we have

$$\operatorname{dist}(T\psi_1, T\psi_2) \leq \sum_{i=0}^{k} \alpha_i \left[q d_i + c \sum_{j=0}^{i-1} d_j \right] \leq \tilde{q} \operatorname{dist}(\psi_1, \psi_2).$$

Therefore the equation $\psi = T\psi$ has a solution $\psi \in B(M_1, \ldots, M_k)$, which completes the proof of Theorem 3.2. □

Now we state solvability conditions for C^∞-functions.

Theorem 3.3. *Let a mapping* $G(x,y) = (F(x), g(x,y))$ *belong to a class* C^∞ *in a neighborhood of a fixed point* (x_0, y_0) *and let the Taylor series of a* C^∞*-function* $\varphi_0(x) = y_0 + \cdots$ *be a formal solution of* (3.0.1) *at* x_0.
Assume that at least one of the following conditions holds:

i) $F'(x_0) \neq 0$ *and the function* $F(F(x)) - x$ *is not flat at* x_0.

ii) The function $F(F(x)) - x$ *is flat at* x_0 *while the functions* $g'_y(x, \varphi_0(x)) - 1$ *and* $g'_y(x, \varphi_0(x)) + 1$ *are not flat at* x_0.

iii) $F'(x_0) = 0$ *while the function* $g'_y(x, \varphi_0(x))$ *is not flat at* x_0.

Then (3.0.1) *has a local* C^∞*-solution whose Taylor series at* x_0 *coincides with* $\hat{\varphi}_0$.

Let us note that this statement contains Theorem 2.7 on the local conjugacy.

Proof of Theorem 3.3. Let us denote by $\mathcal{F}(\epsilon)$ the Frechêt space of C^∞-functions on an interval $I(\epsilon)$ endowed with the norms

$$\|\varphi\|_k = \max_{x \in I(\epsilon)} |\varphi^{(k)}(x)|, \qquad k = 0, 1, \ldots.$$

We are looking for a solution of (3.0.1) represented in the form $\varphi = \psi + \varphi_0$ where $\psi \in C^\infty$ is flat at x_0. It means that instead of (3.0.1) we consider the equation

$$\psi(F(x)) = \tilde{g}(x, \psi(x)) + \gamma(x) \tag{3.1.21}$$

with

$$\tilde{g}(x,y) = g(x,y+\varphi_0(x)) - g(x,\varphi_0(x))$$

and the C^∞-function

$$\gamma(x) = g(x,\varphi_0(x)) - \varphi_0(F(x)),$$

flat at $x = x_0$. Note that

$$\frac{\partial \tilde{g}(x_0,0)}{\partial y} = \frac{\partial g(x_0,y_0)}{\partial y}.$$

Let us assume $x_0 = 0$, $y_0 = 0$ and represent (3.1.21) in the form

$$\psi(F(x)) = a(x)\psi(x) + \psi^2(x)h(x,\psi(x)) + \gamma(x) \tag{3.1.22}$$

where

$$a(x) = \frac{\partial \tilde{g}(x,0)}{\partial y}, \quad h(x,y) = \frac{\tilde{g}(x,y) - a(x)y}{y^2}, \quad |x| \le \varepsilon,\ |y| \le \delta,$$

are C^∞-functions.

We start with the simplest case $\lambda = |F'(0)| \ne 0$, $\mu = |a(0)| \ne 0$.

If $\lambda < 1$, then we write (3.1.22) in the form

$$\psi(x) = \frac{1}{a(x)}\psi(F(x)) - \frac{1}{a(x)}\psi^2(x)h(x,\psi(x)) - \frac{1}{a(x)}\gamma(x).$$

Given an infinite matrix $\{c_{k\nu}\}_{k,\nu=1}^\infty$ and a sequence ν_k, $k = 1,2,\ldots$, the set $B(\{c_{k\nu}\},\{\nu_k\})$ of all C^∞-functions on $I(\varepsilon)$ satisfying

$$|\psi^{(k)}(x)| \le c_{k\nu}|x|^\nu, \quad \nu \ge \nu_k,\ k = 0,1,2,\ldots,$$

is convex and compact in the subspace $\mathcal{F}_0(\epsilon) \subset \mathcal{F}(\epsilon)$ of functions flat at $x = 0$. Let us choose the parameters $c_{k\nu}$ and ν_k such that $B(\{c_{k\nu}\},\{\nu_k\})$ is invariant with respect to the operator

$$T\psi(x) = \frac{1}{a(x)}\psi(F(x)) - \frac{1}{a(x)}\psi^2(x)h(x,\psi(x)) - \frac{1}{a(x)}\gamma(x). \tag{3.1.23}$$

Since $\lambda < 1$, there exists $\varepsilon_0 > 0$ such that

$$|F'(x)| \le q, \quad x \in I(\epsilon), \quad \epsilon \le \epsilon_0,$$

with $q < 1$. Let us set

$$\mu_1 = \max_{|x|\le\varepsilon_0} \frac{1}{|a(x)|},$$

$$\mu_2 = \max_{|x|\le\varepsilon_0,\,|y|\le\varepsilon_0} \left|\frac{h(x,y)}{a(x)}\right|, \quad \mu_3 = \max_{|x|\le\varepsilon_0,\,|y|\le\varepsilon_0} \left|\frac{1}{a(x)}\frac{\partial h(x,y)}{\partial y}\right|,$$

and

$$b_{k\nu} = \max_{|x|\le\varepsilon_0}\left|\frac{\gamma^{(k)}(x)}{a(x)}\right|\frac{1}{|x|^\nu}, \quad k,\nu = 0,1,\dots .$$

We choose $\tilde{q}\in(0,1)$ and fix first ν_0 sufficiently large and afterwards $\varepsilon \le \epsilon_0$ sufficiently small such that

$$\mu_1 q^{\nu_0} + 2\mu_2\epsilon^{\nu_0} + \mu_3\epsilon^{2\nu_0} + b_{0\nu_0+1}\epsilon \le \tilde{q}.$$

If now $|\psi(x)| \le |x|^{\nu_0}$, then

$$|T\psi(x)| \le (\mu_1 q^{\nu_0} + \mu_2|x|\nu_0 + b_{0\nu+1})|x|)|x|^{\nu_0} \le \tilde{q}|x|^{\nu_0},\ x\in I(\varepsilon).$$

Furthermore, if

$$|\psi(x)| \le c_{0\nu}|x|^\nu, \quad \nu \ge \nu_0,$$

then

$$|T\psi(x)| \le ((\mu_1 q^\nu + \mu_2|x|^{\nu_0})c_{0\nu} + b_{0\nu})|x|^\nu \le (\tilde{q}c_{0\nu} + b_{0\nu})|x|^\nu.$$

Hence choosing

$$c_{0\nu_0} = 1, \quad c_{0\nu} > \frac{b_{0\nu}}{1-\tilde{q}}, \qquad \nu > \nu_0,$$

we find that if $|\psi(x)| \le c_{0\nu}|x|^\nu,\ \nu > \nu_0$, then

$$|T\psi(x)| \le \tilde{q}c_{0\nu}|x|^\nu, \quad \nu > \nu_0.$$

Differentiating the function $T\psi(x)$, we obtain

$$(T\psi)^{(k)}(x) = \frac{1}{a(x)}\psi^{(k)}(F(x))(F'(x))^k$$

$$-\frac{2}{a(x)}\psi(x)\psi^{(k)}(x)h(x,\psi(x)) - \frac{1}{a(x)}\psi^2(x)h_y(x,\psi(x))\psi^{(k)}(x)$$

$$+ R_k(x,\psi(x),\psi(F(x)),\dots,\psi^{(k-1)}(x),\psi^{(k-1)}(F(x)))$$

with some C^∞-function $R_k(x,y_1,\dots,y_{2k})$.

If $|\psi^{(k)}(x)| \le c_{k\nu}|x|^\nu$, then

$$|(T\psi)^{(k)}(x)| \le (\mu_1 q^\nu + 2\mu_2|x|^{\nu_0} + \mu_3|x|^{2\nu_0})c_{k\nu}|x|^\nu + r_{k\nu}(c_{0\nu},\dots,c_{k-1,\nu})|x|^\nu$$

with some positive functions $r_{k\nu}(y_0,\dots,y_{k-1})$. Now we set recurrently

$$c_{k\nu} = \frac{r_{k\nu}(c_{0\nu},\dots,c_{k-1,\nu})}{1-\tilde{q}}, \quad k \ge 1,\ \nu \ge \nu_0,$$

and obtain a convex compact set $B(\{c_{k\nu}\},\{\nu_k\}) \subset \mathcal{F}_0(\epsilon)$ with $\nu_k = \nu_0$, which is invariant with respect to the operator T. According to the Schauder–Tichonov theorem [42] there exists a solution ψ of (3.1.22) belonging to this set.

If $\lambda > 1$ and $\mu \neq 0$, then we accept $\psi(F(x))$ as an unknown function in (3.1.22) and arrive at the case we have just considered.

Let now $F'(0) = 1$ and $\mu \neq 0$. Since $\hat{F}(x) \neq x$, then

$$F(x) = x(1 + \alpha x^r + \cdots), \quad \alpha \neq 0, \quad r \geq 1.$$

It is sufficient to prove that there exists a local solution ψ of (3.1.22) for $x \geq 0$, flat at $x = 0$, and hence we will restrict ourselves to estimating $T\psi(x)$ for $x \geq 0$ only. The fixed point $x_0 = 0$ either attracts or repels such x's depending on a sign of α. It is easy to see that for every function $a \in C^\infty$ there exist $\varepsilon > 0$ and $c > 0$ such that either

$$|a(x)| \geq 1 - cx^r, \quad x \in [0, \varepsilon], \tag{3.1.24}$$

or

$$|a(x)| \leq 1 - cx^\rho, \quad x \in [0, \varepsilon], \tag{3.1.25}$$

with some $\rho \in (0, r)$. Indeed, if $|a(0)| > 1$, then (3.1.24) is valid with arbitrary $c > 0$, and if $|a(0)| < 1$, then (3.1.25) is fulfilled also with arbitrary $c > 0$ and sufficiently small $\varepsilon > 0$. If $|a(0)| = 1$ and $a(x) = \pm 1 + b(x)$ with a function b flat at $x = 0$, then again (3.1.24) is valid. The remaining possibility is that $\pm a(x) = 1 + cx^p + o(x^p)$ with some $c \neq 0$ and $p \geq 1$. For $c > 0$ it implies (3.1.24) and for $c < 0$ we choose arbitrary positive $\rho < \min(r, p)$ and obtain (3.1.25).

Assume that $\alpha < 0$ and (3.1.24) is fulfilled. Then

$$|a(x)|^{-1} \leq 1 + c_1 x^r, \; |F(x)| \leq x(1 - c_2 x^r), \; |F'(x)| \leq 1 - c_2 x^r,$$

with some $c_1 > 0$, $c_2 > 0$, for all $x \in [0, \varepsilon]$ and we represent (3.1.22) in the form

$$\psi(x) = T\psi(x)$$

with the same operator T as in (3.1.23).

To construct a T-invariant convex compact subset in $\mathcal{F}_0(\epsilon)$, let $|\psi(x)| \leq x^{\nu_0}$ with some $\nu_0 > 0$ and $x \in [0, \varepsilon]$. Then

$$|T\psi(x)| \leq ((1 + c_1 x^r)(1 - c_2 x^r)^{\nu_0} + \mu_2 x^{\nu_0} + b_{0,\nu_0+r+1} x^{r+1}) x^{\nu_0}.$$

If ν_0 is sufficiently large and $\varepsilon > 0$ is sufficiently small, then

$$(1 + c_1 x^r)(1 - c_2 x^r)^\nu \leq 1 - c_3 x^r, \quad \nu \geq \nu_0, \; x \leq \varepsilon.$$

Hence we can fix ν_0 and ε such that

$$(1 + c_1x^r)(1 - c_2x^r)^\nu + \mu_2x^{\nu_0} + \mu_3x^{\nu_0} + b_{0,2\nu_0}x^{\nu_0} \le 1 - \theta x^r$$

with some $\theta > 0$ for all $x \in [0,\epsilon]$. Then the estimate $|\psi(x)| \le x^{\nu_0}$ implies $|T\psi(x)| \le x^{\nu_0}$, while the estimate $|\psi(x)| \le c_{0,\nu}x^\nu$, $\nu \ge \nu_0$, yields

$$|T\psi(x)| \le ((1 - \theta x^r)c_{0,\nu} + b_{0,\nu+2r}x^r)x^\nu.$$

Choosing

$$c_{0,\nu_o} = 1, \qquad c_{0,\nu} > \frac{b_{0,\nu+2r}}{\theta}, \qquad \nu > \nu_0,$$

we obtain

$$|T\psi(x)| \le c_{0,\nu}x^\nu, \qquad \nu \ge \nu_0.$$

Furthermore, if $|\psi^{(k)}(x)| \le c_{k\nu}x^\nu$, $k \ge 1$, then

$$|(T\psi)^{(k)}(x)| \le ((1 + c_1x^r)(1 - c_2x^r)^{\nu+k} + \mu_3x^{\nu_0})c_{k\nu}x^\nu$$

$$+ R_k(c_{0,\nu+r}, \ldots, c_{k-1,\nu+r})x^{\nu+r}.$$

Therefore if ν_k is sufficiently large, then

$$|(T\psi)^{(k)}(x)| \le \left(1 - \theta x^r + \frac{R_k(c_{0,\nu+r}, \ldots, c_{k-1,\nu+r})}{c_{k\nu}}x^r\right) c_{k\nu}x^\nu$$

for $\nu \ge \nu_k$. Choosing recurrently

$$c_{k\nu} > \frac{R_k(c_{0,\nu+r}, \ldots, c_{k-1,\nu+r})}{\theta},$$

we have

$$|(T\psi)^{(k)}(x)| \le c_{k\nu}x^\nu, \quad \nu \ge \nu_k.$$

With the above choice of parameters the set $B(\{c_{k,\nu}\}, \{\nu_k\}) \subset \mathcal{F}_0(\epsilon)$ is T-invariant and hence there exists a solution of the equation $\psi = T\psi$.

The case $|a(x)| \le 1 - cx^\rho$, $\rho < r$, corresponding to (3.1.25) is technically more complicated. In this case instead of (3.1.23) we consider the equivalent equation

$$\psi(x) = a(F^{-1}(x))\psi(F^{-1}(x))$$

$$+ \psi^2(F^{-1}(x))h(F^{-1}(x), \psi(F^{-1}(x))) + \gamma(F^{-1}(x)). \tag{3.1.26}$$

To prove that it has a local solution $\psi(x)$, $x \in [0,\varepsilon]$, with some $\varepsilon > 0$, we introduce the space $\mathcal{F}_1(\epsilon)$ of all C^∞-functions on $\mathbb{R}$ vanishing outside the interval $[0,\varepsilon]$. Denote by $\tilde{a}(x)$ and $\tilde{\gamma}(x)$ some functions from $\mathcal{F}_1(\epsilon)$ coinciding with $a(F^{-1}(x))$ and $\gamma(F^{-1}(x))$, respectively, in $[0,2\varepsilon]$. Then the operator

$$\tilde{T}\psi(x) = \begin{cases} \tilde{a}(x)\psi(F^{-1}(x)) \\ \quad + \psi^2(F^{-1}(x))h(F^{-1}(x),\psi(F^{-1}(x)) + \tilde{\gamma}(x), & x \in [0,\varepsilon], \\ 0, & x \geq \varepsilon, \end{cases}$$

is well defined on the set $\{\, \psi \in \mathcal{F}_1(\epsilon) : |\psi(x)| \leq \delta \,\}$, maps it into $\mathcal{F}_1(\epsilon)$ and the function $\tilde{a}$ satisfies (3.1.25).

To construct a $\tilde{T}$-invariant convex compact subset in $\mathcal{F}_1(\epsilon)$ we set

$$\mu = \max_{\substack{0 \leq x \leq \varepsilon \\ |y| \leq \delta}} \left| h(F^{-1}(x),y)\right| + \left| h_y(F^{-1}(x),y)\right|, \quad b_{k\nu} = \max_{0\leq x\leq\varepsilon} \left| \frac{\tilde{\gamma}^{(k)}(x)}{x^\nu} \right|.$$

Since $\rho < r$ and representations

$$F^{-1}(x) = x(1 + c_1 x^r + o(x^r)), \quad (F^{-1}(x))' = 1 + c_2 x^r + o(x^r)$$

are valid, we can choose $\varepsilon > 0$ and $\sigma > \rho$ such that

$$1 - cx^\rho + (3\mu + 1)x^\sigma \leq x^\sigma, \quad 0 \leq x \leq \epsilon.$$

If now

$$\mathcal{K}_0 = \{\, \psi \in \mathcal{F}_1(\epsilon) : |\psi(x)| \leq x^\sigma \,\},$$

then

$$|\tilde{T}\psi(x)| \leq (1 - cx^\rho)(F^{-1}(x))^\sigma + \mu(F^{-1}(x))^{2\sigma} + b_{0,2\sigma}x^{2\sigma} \leq x^\sigma, \psi \in K_0,$$

which shows that $\mathcal{K}_0$ is invariant with respect to $\tilde{T}$.

Furthermore, for every $\nu \geq \sigma$ and $k \geq 0$ we define functions

$$\beta_{k\nu}(x) = \begin{cases} x^\nu, & 0 \leq x \leq \varepsilon_{k\nu}, \\ \alpha_{k\nu}x^{-\nu}, & x \geq \varepsilon_{k\nu}, \end{cases}$$

with some numbers $\alpha_{k\nu} > 0$ and $\varepsilon_{k\nu} < \varepsilon$.

To choose these numbers we compare the functions

$$\omega_k(x) = (1 - cx^\rho)((F^{-1}(x))')^k + (3\mu + 1)x^\sigma = 1 - cx^\rho + o(x^\rho),$$

with functions

$$\frac{\beta_{k\nu}(x)}{\beta_{k\nu}(F^{-1}(x))} = \begin{cases} \left(\frac{x}{F^{-1}(x)}\right)^\nu, & 0 \leq x \leq F^{-1}(x) \leq \varepsilon_{k\nu}, \\ \frac{(xF^{-1}(x))^\nu}{\alpha_{k,\nu}}, & 0 \leq x \leq \varepsilon_{k\nu} \leq F^{-1}(x), \\ \left(\frac{F^{-1}(x)}{x}\right)^\nu, & \varepsilon_{k\nu} \leq x \leq F^{-1}(x). \end{cases}$$

Since $F^{-1}(x) = x(1 + O(x^r))$ and $\rho < r$, we can fix $\varepsilon_{k\nu} > 0$ small enough for the inequalities

$$\omega_k(x) \le \frac{\beta_{k\nu}(x)}{\beta_{k\nu}(F^{-1}(x))} \tag{3.1.27}$$

to be valid for all $k \ge 0$, $\nu \ge \sigma$ and $0 \le x \le F^{-1}(x) \le \varepsilon_{k\nu}$. The condition $F^{-1}(x) \ge \varepsilon_{k\nu}$ implies $xF^{-1}(x) \ge F(\varepsilon_{k\nu})\varepsilon_{k\nu}$ and we can fix $\alpha_{k\nu} > 0$ such that (3.1.27) is fulfilled for all k, $\nu \ge \sigma$ and $0 \le x \le \varepsilon_{k\nu} \le F^{-1}(x)$. At last, since $F^{-1}(x)x^{-1} = 1 + c_1 x^r + o(x^r)$ with $c_1 > 0$, for every fixed $k \ge 0$ there exists $\nu_k > \nu_{k-1} \ge \sigma$ such that (3.1.27) is valid for all k, $\nu \ge \nu_k$ and $x \in [0, \varepsilon]$.

With $\alpha_{k\nu}$ and $\varepsilon_{k\nu}$ being fixed, there exist numbers $A_{k\nu}$, $B_{k\nu}$, $C_{k\nu}$, $D_{k\nu}$ such that

$$B_{k\nu}x^\nu \le \beta_{k\nu}(x) \le A_{k\nu}x^\nu, \quad D_{k\nu}x^\nu \le \beta_{k\nu}(F^{-1}(x)) \le C_{k\nu}x^\nu.$$

Our next step is to choose numbers $c_{k\nu}$ for the compact set

$$\mathcal{K} = \{ \psi \in: |\psi^{(k)}(x)| \le c_{k\nu}\beta_{k\nu}(x), \quad k = 0, 1, 2, \ldots ; \ \nu \ge \nu_k \}$$

to be $\tilde{T}$-invariant. To this end we note that if, in addition to the condition $\psi \in \mathcal{K}_0$, we have $|\psi(x)| \le c_{0\nu}\beta_{0\nu}(x)$, $\nu > \nu_0 = \sigma$, then

$$|\tilde{T}\psi(x)| \le ((1 - cx^\rho + \mu x^\sigma)c_{0\nu} + b_{02\nu}x^\nu D_{0\nu}^{-1})\beta_{0\nu}(F^{-1}(x)).$$

Therefore with $\nu \ge \sigma$ and $c_{0\nu} \ge b_{0,2\nu}D_{0\nu}^{-1}$ we obtain

$$|\tilde{T}\psi(x)| \le c_{0\nu}\beta_{0\nu}(x), \quad \nu \ge \sigma.$$

For derivatives of $\tilde{T}\psi(x)$ we have

$$(\tilde{T}\psi)^{(k)}(x) = \{ \tilde{a}(x) + 2\psi(F^{-1}(x))h(F^{-1}(x), \psi(F^{-1}(x)))$$

$$+ \psi^2(F^{-1}(x))h_y(F^{-1}(x)), \psi(F^{-1}(x))\}\psi^{(k)}(F^{-1}(x))((F^{-1}(x))')^k$$

$$+ R_k(x, \psi(F^{-1}(x)), \ldots, \psi^{(k-1)}(F^{-1}(x)))$$

with some C^∞-functions $R_k(x, y_0, \ldots, y_{k-1})$. Now an estimate follows;

$$|(\tilde{T}\psi)^{(k)}(x)| \le (1 - cx^\rho + 3\mu x^{\nu_0})c_{k\nu}\beta_{k\nu}(F^{-1}(x))$$

$$+ r_k(c_{0,2\nu}, \ldots, c_{k-1,2\nu})|x|^{2\nu}, \quad \nu \ge \nu_k, \quad k \ge 1,$$

and we can choose $c_{k\nu}$ recurrently such that the estimate

$$|(\tilde{T}\psi)^{(k)}(x)| \le c_{k\nu}\beta_{k\nu}(x), \ \nu \ge \nu_k, \ k \ge 0,$$

is valid. Therefore the set $\mathcal{K}$ is $\tilde{T}$-invariant and equation (3.1.22) has a solution $\psi \in \mathcal{F}_1(\epsilon)$.

If $F'(0) \neq 0$ and $a(0) = 0$, then instead of (3.1.23) we consider (3.1.26). Now

$$|a(F^{-1}(x))| \leq cx, \quad F^{-1}(x) = \lambda x + o(x), \quad \lambda > 0,$$

and in the case $\lambda \geq 1$, we set

$$\omega_k(x) = cx((F^{-1}(x))')^k + (3\mu + 1)x^\sigma, \quad 0 \leq x \leq \varepsilon, \quad k = 0, 1, \ldots,$$

choose ε, $\varepsilon_{k\nu}$ and $\alpha_{k\nu}$ to satisfy (3.1.27) and proceed as above.

The case $0 < \lambda < 1$, $a(0) = 0$, is much simpler and was in fact treated at the beginning of the proof.

The previous arguments prove Theorem 3.3 if F preserves the orientation and conditions *i)* are satisfied.

Let again F preserve the orientation and conditions *ii)* be satisfied. Then we can assume $F(x) = x + o(x^r)$ and $|a(x)| = 1 + cx^\rho + o(x^\rho)$ with some $c \neq 0$ and $\rho \in [0, r)$. Now depending on the sign of c, either (3.1.24) or (3.1.25) is fulfilled and we can proceed with the case $F'(0) > 0$ in the same way as above.

Let us now consider the case $F'(0) < 0$. If either condition *i)* or *ii)* is fulfilled, then as we have already proved there exists a C^∞-solution ψ of the iterated equation

$$\psi(F^2(x)) = \tilde{g}(F(x), \tilde{g}(x, \psi(x)) + \gamma(x)) + \gamma(F(x)),$$

flat at $x = 0$. It is easy to see that

$$\varphi(x) = \begin{cases} \tilde{g}(F^{-1}x, \psi(F^{-1}(x))) + \gamma(F^{-1}(x)), & x \leq 0, \\ \psi(x), & x \geq 0, \end{cases}$$

is a well-defined C^∞-solution of (3.1.22). This completes the proof of Theorem 3.3 under conditions *i)* and *ii)*.

For the remaining case $F'(0) = a(0) = 0$ we note that the function $a(x)$ is not flat at $x = 0$ and therefore $|a(x)| \geq c_1|x|^r$, $0 \leq x \leq \epsilon_0$, $r > 0$. As before, the operator (3.1.23) is well defined in the subspace $\mathcal{F}_0(\epsilon)$ of functions flat at $x = 0$. Moreover, since $|F(x)| \leq c_2|x|^2$, $|F'(x)| \leq c|x|$, the inequality $|\psi^{(k)}(x)| \leq c_{k\nu}|x|^\nu$ implies

$$\frac{1}{|a(x)|}|\psi^{(k)}(F(x))||F'(x)|^k \leq c_2^\nu c_3^k c_1^{-1} c_{k\nu}|x|^{2\nu+k-r}.$$

We set

$$\mu = \max_{0\leq x\leq\epsilon_0} \frac{2x^r}{|a(x)|} + \max_{0\leq x\leq\epsilon_0;\ |y|\leq\delta_0} |h(x,y)| + \max_{0\leq x\leq\epsilon_0;\ |y|\leq\delta_0} |h_y'(x,y)|,$$

and, as for $\lambda < 1$, arrive at the estimate

$$|T\psi^{(k)}(x)| \leq (c_2^\nu c_3^k c_1^{-1} x^{\nu+k-r} + \mu x) c_{k\nu} x^\nu + r_k(c_{0,\nu}, \ldots, c_{k-1,\nu}) x^\nu$$

where $0 \leq x \leq \epsilon \leq \epsilon_0$. With a proper choice of parameters ϵ, ν_k and $c_{k\nu}$ we obtain a T-invariant subset $B(\{c_{k\nu}, \nu_k)\}$ and a solution ψ of equation (3.1.23). $\square$

The following examples show that conditions of Theorem 3.3 are not redundant.

Example 3.4. Consider the equation

$$\varphi(x) - \varphi(F(x)) = \tau(x), \quad F(x) = x(1 - \delta(x))$$

with

$$\delta(x) = \begin{cases} e^{-x^{-4}}, & x \neq 0, \\ 0, & x = 0, \end{cases} \qquad \tau(x) = \begin{cases} e^{-x^{-2}}, & x \neq 0, \\ 0, & x = 0. \end{cases}$$

Both functions δ and τ are flat at $x = 0$, $\hat{F}(x) = x$ and the trivial series $\hat{\varphi}_0 = 0$ is a formal solution. Let us set $x_k = (\ln k)^{-1/2}$, $k = 2, 3, \dots$. It is easy to check that $F(x_k) \geq x_{n+k}$ for all sufficiently large k and $n = 1, 2, \dots$, and if φ is a solution, then

$$\varphi(x_k) = \varphi(F^n(x_k)) + \sum_{j=0}^{n-1} \tau(F^j(x_k)), \quad k = 2, 3, \dots, \; n = 1, 2, \dots.$$

Since $F^n(x_k) \to 0$ as $n \to \infty$ and

$$\sum_{j=0}^{n-1} \tau(F^j(x_k)) \geq \sum_{j=1}^{n-1} \tau(x_{k+j}) \to \infty$$

as $n \to \infty$, a solution φ cannot be bounded at $x = 0$.

Example 3.5. Consider the equation

$$\varphi(x^2) = x(\varphi^2(x) + \tau(x))$$

with the same function τ as in the previous example.

Here $F(x) = x^2$, $F'(0) = 0$, the trivial series $\hat{\varphi}_0 = 0$ is a formal solution as well, however the equation has no continuous solutions: the function $\varphi(x^2)$ is even, while the expression in the right hand side changes its sign after the substitution $x \to -x$.

Theorem 3.4. *Let the mapping* $G(x,y) = (F(x), g(x,y))$ *be analytic in a neighborhood of its fixed point* (x_0, y_0), *let* $\hat{\varphi}_0 = y_0 + c_1(x - x_0) + \cdots$ *be a formal solution of* (3.0.1) *at* x_0, *and let at least one of the following assumptions hold:*

i) $|F'(x_0)| > 1$;

ii) $|F'(x_0)| < 1$, $g_y(x_0, y_0) \neq 0$;

iii) $F'(x_0) = 0$ *and the formal series* $g_y(x, \hat{\varphi}_0(x))$ *at* x_0 *is not zero.*

Then for every $p = 0, 1, 2, \dots$ *there exists a local analytic solution of* (3.0.1) *whose Taylor polynomial of degree* p *at* x_0 *coincides with the corresponding segment of the formal solution* $\hat{\varphi}_0$.

Proof. Without loss of generality we assume $x_0 = y_0 = 0$ and $|F'(0)| < 1$; if the opposite inequality holds, we use the substitution $x \to F^{-1}(x)$.

We fix a number $p_0 \geq 1$ such that

$$|g'_y(0,0)||\lambda|^{p_0} < 1$$

and, given an integer $p > p_0$, set $\varphi = \psi + \varphi_1$ where φ_1 is the polynomial segment of degree $p+1$ of the formal solution $\hat{\varphi}_0$. It is sufficient to prove the existence of a local analytic solution ψ of (3.1.22) with

$$\tilde{g}(x,y) = g(x, y + \varphi_1(x)) - g(x, \varphi_1(x))$$

and

$$\gamma(x) = g(x, \varphi_1(x)) - \varphi_1(F(x)),$$

satisfying conditions

$$\psi^{(k)}(0) = 0, \quad 0 \leq k \leq p.$$

Note that γ does satisfy the same conditions. If as above $a(z) = \tilde{g}'_y(z,0)$ and $a(0) \neq 0$, then we can write (3.1.22) in the form

$$\psi(x) = T\psi(x)$$

where T is the same as in (3.1.23). With a proper choice of $\varepsilon > 0$ the operator T is well defined in the space $A(\varepsilon)$ of functions ψ analytic in the disc $\{ z \in \mathbb{C} : |z| < \varepsilon \}$ endowed with the topology of uniform convergence on compact subsets. It is sufficient to construct a T-invariant convex compact subset in $A(\varepsilon)$.

To this end we fix $\varepsilon_0 > 0$ such that

$$|a(z)|^{-1}|z^{-1}F(z)|^{p_0} \leq q < 1, \quad z \in \mathbb{C}, \ |z| \leq \varepsilon_0,$$

with some number q and set

$$\mu = \max_{|z| \leq \varepsilon_0, \ |u| \leq \varepsilon_0,} \frac{|h(z,u)|}{|a(z)|}, \qquad b = \max_{|z| \leq \varepsilon_0} \frac{|\gamma(z)|}{|a(z)||z|^{p+1}}.$$

At last, we fix $\epsilon < \epsilon_0$ such that

$$q + \mu\epsilon^p + b\epsilon < 1.$$

If $|\psi(z)| \leq |z|^p, \ |z| < \varepsilon$, then

$$|T\psi(z)| \leq q|z|^p + \mu|z|^{2p} + b|z|^{p+1} \leq |z|^p.$$

In other words, the convex compact subset

$$\{ \psi \in A(\varepsilon) : |\psi(z)| \leq |z|^p, \ |z| < \varepsilon \}$$

is T-invariant, completing the proof in the case $a(0) \neq 0$.

In case *iii)* we have $|F(z)| \leq c|z|^2$ and $|a(z)| \geq c|z|^r$ with some $r > 0$. It leads us to an estimate

$$|T\psi(z)| \leq c|z|^{2p-r} + \mu|z|^{2p-r} + b|z|^{p+1}, \qquad |z| < \epsilon,$$

with some parameters c, μ, b for all ψ satisfying $|\psi| \leq |z|^p, |z| < \epsilon$, and with a proper choice of ϵ we obtain a solution of (3.1.23). □

Conditions of the local analytic solvability in Theorem 3.4 are more restrictive than in the C^∞-case. We will see in the next section that its statement may fail if these conditions are not satisfied.

3.1.4 More on analytic solutions

In contrast with the case of C^∞-functions, Theorem 3.4 does not cover the cases $F'(x_0) = 1$ and

$$0 < |F'(x_0)| < 1, \quad \frac{\partial g(x_0, y_0)}{\partial y} = 0.$$

One of the underlying reasons is that in the space of local analytic functions there exist linear functionals which are not finite linear combinations of derivatives at a fixed point and which may be obstacles to the local solvability. To demonstrate this phenomenon we consider several examples.

Example 3.6. If $|\mu| < 1$, then the equation

$$\varphi(x) - \mu\,\varphi\left(\frac{x}{(1+x^2)^{1/2}}\right) = \gamma(x)$$

has the unique C^k-solution

$$\varphi_0(x) = \sum_{n=0}^{\infty} \mu^n \gamma\left(\frac{x}{(1 + nx^2)^{1/2}}\right)$$

for every C^k-function γ with either finite $k \geq 0$ or $k = \infty$. Let us show that the situation is different if $k = \aleph$.

If γ is analytic in a neighborhood of $x = 0$, then φ_0 is analytic in an open interval $(0, \epsilon)$ for sufficiently small $\epsilon > 0$. Nevertheless, the equation may have no local analytic solutions at $x = 0$ for some analytic functions γ. For example, the function

$$\varphi_0(z) = \sum_{n=0}^{\infty} \mu^n \frac{z}{(1 + nz^2)^{1/2}}$$

is analytic in $\mathbb{C}$ except poles at $z_n = \pm i n^{-1/2}$, $n = 1, 2, \ldots$, and the essential singularity at $z = 0$. If φ is a local analytic solution at $x = 0$ corresponding to

$\gamma(x) = x$, then the function $\psi = \varphi - \varphi_0$ is continuous in $[0, \epsilon]$ and satisfies the homogeneous equation

$$\psi(x) - \mu\, \psi\left(\frac{x}{(1 + x^2)^{1/2}}\right) = 0.$$

Iterating we obtain $\psi(x) \equiv 0$ and $\varphi \equiv \varphi_0$. On the other side, the singular points of φ_0 are not removable and hence a local analytic solution φ does not exist.

Example 3.7. The equation

$$\varphi(x) - \varphi\left(\frac{x}{x + 1}\right) = \gamma(x), \tag{3.1.28}$$

is formally solvable for every $\gamma \in C^\infty$ such that $\gamma(0) = \gamma'(0) = 0$. It follows from Theorem 3.3 that there exists a local C^∞-solution at $x_0 = 0$, but Theorem 3.4 is not applicable, since here $F(x) = x(1 + x)^{-1}$ and $F'(0) = 1$.

If $\gamma(0) = \gamma'(0) = 0$, then the series

$$\varphi(x) = \sum_{n=0}^{\infty} \gamma\left(\frac{x}{1 + nx}\right)$$

converges for $x > 0$ and is an analytic solution of (3.1.28) in $(0, \varepsilon)$ where ε is the radius of analyticity of γ. However generally speaking, it is not analytic in a neighborhood of $x = 0$.

Indeed, the domain D^+ bounded by the curves

$$\{\, z \in \mathbb{C} : |z| = \epsilon, \ \Re z \geq 0 \,\},$$

and

$$\{\, z \in \mathbb{C} : \left|z - \frac{i\varepsilon}{2}\right| = \frac{\varepsilon^2}{4}, \ \Re z \leq 0\}, \quad \{\, z \in \mathbb{C} : \left|z + \frac{i\varepsilon}{2}\right| = \frac{\varepsilon^2}{4}, \ \Re z \leq 0\},$$

is invariant with respect to the mapping $G(z) = z(1 + z)^{-1}$, while its mirror reflection $D^- = -D^+$ is invariant with respect to the inverse mapping $G(z) = z(1 - z)^{-1}$. The series

$$\varphi_+(z) = \sum_{n=0}^{\infty} \gamma\left(\frac{z}{1 + nz}\right), \ \varphi_-(z) = -\sum_{n=1}^{\infty} \gamma\left(\frac{z}{1 - nz}\right) \tag{3.1.29}$$

converge in D^+ and D^-, respectively, to analytic solutions of (3.1.28). In addition, since

$$\left|\gamma\left(\frac{x}{1 + nx}\right)\right| \leq \frac{c}{n^2}, \qquad x \in [0, \epsilon],$$

the former series converges uniformly in the segment $[0, \epsilon]$ and defines a continuous function $\varphi_+(x)$ vanishing at $x = 0$. Similarly, $\varphi_-(x)$ is continuous in $[-\epsilon, 0]$ and vanishes at $x = 0$ as well.

Assume now that φ is a solution of (3.1.28) analytic in a neighborhood of $z = 0$. Then the functions $\psi_\pm(z) = \varphi(z) - \varphi_\pm(z)$ are analytic in $D^\pm$, respectively, and satisfy the homogeneous equations $\psi_\pm(z) = \psi_\pm(F(z))$ in the corresponding domains. Iterating these equations in the respective intervals of the real axis, we find that the functions $\psi_+(x)$ and $\psi_-(x)$ are constant in $[0, \epsilon]$ and $[-\epsilon, 0]$, implying that $\psi_+(z)$ and $\psi_-(z)$ are constant in D^+ and D^-, respectively.

Since $\varphi_\pm(0) = 0$, we conclude that $\psi_+(0) = \psi_-(0) = \varphi(0)$ and

$$\sum_{n=-\infty}^{\infty} \gamma\left(\frac{z}{1 + nz}\right) = 0, \quad z \in D^+ \bigcap D^-. \tag{3.1.30}$$

This identity is a necessary condition for the local solvability of (3.1.28). On the other hand, if it is fulfilled, then the function

$$\varphi(z) = \begin{cases} \varphi_+(z), & z \in D^+, \\ \varphi_-(z), & z \in D^-, \end{cases}$$

where φ_+ and φ_- are functions from (3.1.29), is a well-defined local analytic solution of (3.1.28).

Hence the conditions $\gamma(0) = \gamma'(0) = 0$ together with (3.1.30) are necessary and sufficient for the local analytic solvability of (3.1.28).

To express (3.1.30) in terms of Taylor coefficients of γ, we note that the function

$$\theta(z) = \sum_{n=-\infty}^{\infty} \gamma\left(\frac{1}{z + n}\right)$$

with $\gamma(0) = \gamma'(0) = 0$ is analytic and 1-periodic in the domain $\{ z \in \mathbb{C} : |\mathrm{Im}\, z| > \varepsilon^{-1} \}$. The condition (3.1.30) means that θ is identically zero. Hence all its Fourier coefficients vanish and

$$\int_{-\infty}^{\infty} \gamma\left(\frac{1}{t + ic}\right) e^{2ik\pi t} dt = 0, \quad |c| > \varepsilon^{-1}, \ k \in \mathbb{Z}.$$

Substituting here the Taylor expansion of γ we find that (3.1.28) has a local analytic solution if and only if

$$\gamma(0) = \gamma'(0) = 0, \quad \sum_{n=2}^{\infty} \frac{\gamma^{(n)}(0)}{n!} \frac{(2\pi ik)^{n-1}}{(n - 1)!} = 0, \quad k \in \mathbb{Z}.$$

Example 3.8. Consider the equation

$$\varphi(\lambda x) - x\varphi(x) = \gamma(x), \quad \lambda \in (0,1). \tag{3.1.31}$$

Here $F(x) = \lambda x$, $g(x,y) = xy$, $g_y(0,0) = 0$, and Theorem 3.4 may not be applied. It follows from Theorems 3.2 and 3.3 that for every $\gamma \in C^k$, $k \leq \infty$, the equation has a local C^k-solution. On the other hand, if γ is analytic, then it may have not analytic solutions. For instance, if $\gamma(x) = 1$, then the unique formal solution

$$\varphi(x) = \sum_{k=0}^{\infty} \lambda^{-\frac{k(k+1)}{2}} x^k$$

diverges at $x \neq 0$.

More generally, the unique formal solution of (3.1.31) is

$$\hat{\varphi}(x) = \sum_{m=0}^{\infty} \varphi_m x^m$$

with

$$\varphi_m = \lambda^{-\frac{m(m+1)}{2}} \sum_{k=0}^{m} \frac{\gamma^{(k)}(0)}{k!} \lambda^{\frac{k(k-1)}{2}}.$$

Since $|\gamma^{(k)}(0)| \leq ck!r^k$, the series

$$l[\gamma] = \sum_{k=0}^{\infty} \frac{\gamma^{(k)}(0)}{k!} \lambda^{\frac{k(k-1)}{2}}$$

converges. If $l[\gamma] \neq 0$, then the series $\hat{\varphi}(x)$ diverges at $x \neq 0$, and (3.1.31) has no local analytic solutions. On the other hand, if $l[\gamma] = 0$, then the numbers

$$\varphi_m = -\lambda^{-\frac{m(m-1)}{2}} \sum_{k=m+1}^{\infty} \frac{\gamma^{(k)}(0)}{k!} \lambda^{\frac{k(k-1)}{2}}$$

satisfy the estimate $|\varphi_m| \leq \rho^m$ with some $\rho > 0$ and $\hat{\varphi}(x)$ defines a local analytic solution of (3.1.31). Thus equation (3.1.31) has a local analytic solution if and only if

$$\sum_{k=0}^{\infty} \frac{\gamma^{(k)}(0)}{k!} \lambda^{\frac{k(k-1)}{2}} = 0.$$

Let us note that an analytic solution of (3.1.31), if it exists, is unique. In contrast to this, the C^∞-function

$$\psi(x) = \begin{cases} x^{-\frac{1}{2}} \tau_+ \left(\dfrac{\ln x}{\ln \lambda}\right) \exp \dfrac{\ln^2 x}{2 \ln \lambda}, & x > 0, \\ 0, & x = 0, \\ |x|^{-\frac{1}{2}} \tau_- \left(\dfrac{\ln |x|}{\ln \lambda}\right) \exp \dfrac{\ln^2 |x|}{2 \ln \lambda}, & x < 0, \end{cases}$$

with arbitrary 1-periodic C^∞-functions $\tau_\pm(t)$ is a solution of the homogeneous equation $\psi(\lambda x) = x\psi(x)$.

3.2 Global solutions of equations with not more than one fixed point

In this section we study a global solvability problem for equation (3.0.1) under assumptions that F has one fixed point at most and the mapping $G(x,y) = (F(x), g(x,y))$ is a C^k-diffeomorphism of $M \times \mathbb{R}$.

3.2.1 Equations with fixed-point free mappings F

The following proposition is a global version of Theorem 3.1 for $k \le \infty$.

Theorem 3.5. *Let $G(x,y) = (F(x), g(x,y))$ be a C^k-diffeomorphism of the plane $\mathbb{R}^2$, $k \le \infty$, and let $\mathrm{Fix}F = \emptyset$. Then there exists a C^k-diffeomorphism $\Phi : \mathbb{R}^2 \to \mathbb{R}^2$ of the form (3.0.2) which conjugates G with either $G_0(x,y) = (x+1, y)$ or $G_0(x,y) = (x+1, -y)$ depending on the sign of $g'_y(x,y)$.*

Proof. According to Corollary 2.2 there exists a C^k-diffeomorphism $H : \mathbb{R} \to \mathbb{R}$ such that

$$H^{-1}FH(x) = x + 1,$$

and we can assume that

$$G(x,y) = (x+1, g(x,y)).$$

It remains to find a C^k-transformation $\Phi(x,y) = (x, T(x,y))$ such that

$$\Phi G_0 = G\Phi \tag{3.2.1}$$

with $G_0(x,y) = (x+1, \pm y)$. It is easy to see that $T(x,y)$ is a solution of equation

$$g(x, T(x,y)) = T(x+1, \pm y), \quad (x,y) \in \mathbb{R}^2. \tag{3.2.2}$$

To construct such a solution, let $\tau : \mathbb{R} \to [0,1]$ be a C^∞-function vanishing in a neighborhood of $x = 0$ and equal to 1 in a neighborhood of $x = 1$. Set

$$T_0(x,y) = (1 - \tau(x))y + \tau(x)g((x-1)\tau(x), \pm y),$$

according to the sign of $g'_y(x,y)$. If $|x|$ is small, then $\tau(x+1) = 1$ and

$$T_0(x+1, \pm y) = g(x,y).$$

Hence the equality

$$g(x, T_0(x,y)) = T_0(x+1, \pm y)$$

holds in a neighborhood of the point $x = 0$.

Since

$$\frac{\partial T_0(x,y)}{\partial y} = 1 - \tau(x) \pm \tau(x)\frac{\partial g((x-1)\tau(x), \pm y)}{\partial y} > 0,$$

the mapping $\Phi_0(x,y) = (x, T_0(x,y))$ is a C^k-diffeomorphism of the strip $[0,1]\times\mathbb{R} \subset \mathbb{R}^2$. To continue T_0 as a C^k-function to the entire plane we note that according to (3.2.1) the inverse mapping

$$H(x,y) = G^{-1}(x,y) = (x-1, h(x,y))$$

satisfies the equation $G^{-1}\Phi = \Phi G_0^{-1}$ which is equivalent to

$$h(x, T(x,y)) = T(x-1, \pm y). \tag{3.2.3}$$

We use (3.2.2) and (3.2.3) to define recursively

$$T_n(x,y) = g(x-1, T_{n-1}(x-1, \pm y), \quad n \le x \le n+1,\ n = 1, 2, \ldots$$

and

$$T_n(x,y) = h(x + T_{n+1}(x+1, \pm y), \quad n \le x \le n+1,\ n = -1, -2, \ldots .$$

Now

$$T(x,y) = T_n(x,y), \quad n \le x \le n+1,\ n \in \mathbb{Z},$$

is a well-defined C^k-function on the plane $\mathbb{R}^2$ which satisfies (3.2.2) and hence globally conjugates G and G_0. □

Theorem 3.5 leads us to the following description of solutions of equation (3.0.1).

Corollary 3.2. *Let assumptions of Theorem* 3.5 *hold. Then there exists a* C^k-*transformation* (3.0.2) *such that every* C^k-*solution of* (3.0.1) *has the form*

$$\varphi(x) = T(H^{-1}(x), \psi(H^{-1}(x)))$$

with a C^k-*periodic (or anti-periodic) function* ψ. *Conversely, every function* φ *of such a form is a* C^k-*solution of* (3.0.1).

The transformation T_0 which was constructed to prove Theorem 3.5 may not be used for analytic functions. Nevertheless, a global solution of equation (3.0.1) in the analytic case exists which is proved in the following proposition.

Theorem 3.6. *Let* $G(x, y) = (F(x), g(x, y))$ *be an analytic diffeomorphism, and let* $\operatorname{Fix} F = \emptyset$. *Then equation* (3.0.1) *has an analytic solution on the entire real line.*

Proof. By virtue of Corollaries 2.2 and 3.1 we can assume that $F(x) = x + 1$ and that there exists a local analytic solution φ_0 of (3.0.1) defined in a neighborhood of $[0, 1]$. It means that φ_0 is analytic in a neighborhood of $[0, 1]$ and

$$\varphi_0(x + 1) = g(x, \varphi_0(x))$$

for x from a neighborhood of $x = 0$. Let, as above, $H(x, y) = (x - 1, h(x, y))$ be the inverse mapping to $G(x, y)$. Set

$$\varphi_n(x) = g(x - 1, \varphi_{n-1}(x - 1)), \quad n \leq x \leq n + 1, \; n = 1, 2, \ldots$$

and

$$\varphi_n(x) = h(x + 1, \varphi_{n+1}(x + 1)), \quad n \leq x \leq n + 1, \; n = 0, -1, -2, \ldots .$$

Then

$$\varphi(x) = \varphi_n(x), \quad n \leq x \leq n + 1, \; n \in \mathbb{Z},$$

is a well-defined analytic solution of (3.0.1). □

If $g(x, y) = a(x)y + \gamma(x)$, then (3.0.1) is the linear equation

$$\varphi(F(x)) = a(x)\varphi(x) + \gamma(x). \tag{3.2.4}$$

The invertibility condition in this case means that F is a diffeomorphism and $a(x) \neq 0$, $x \in \mathbb{R}$, and we are looking for transformation (3.0.2) in the form

$$\Phi(x, y) = (H(x), S(x)y + T(x)). \tag{3.2.5}$$

We assume $F(x) = x + 1$ and set $H(x) = x$. The functions $S(x)$ and $T(x)$ are solutions of the equations

$$S(x + 1) = |a(x)|S(x), \quad T(x + 1) = a(x)T(x) + \gamma(x), \tag{3.2.6}$$

respectively, while the function $\tau(x) = \ln S(x)$ satisfies the simplest cohomological equation

$$\tau(x + 1) = \tau(x) + b(x), \quad b(x) = \ln |a(x)|.$$

Since Theorems 3.5 and 3.6 cover all smoothness classes, including analytic functions, equations for T and τ are globally C^k-solvable and we arrive at the following proposition.

Corollary 3.3. *Every linear equation* (3.2.4) *with* C^k*-functions* $a(x) \neq 0$ *and* $\gamma(x)$ *and a* C^k*-diffeomorphism* $F : \mathbb{R} \to \mathbb{R}$, $\operatorname{Fix} F = \emptyset$, *may be reduced to the form*

$$\psi(x + 1) = \pm\psi(x).$$

using a linear C^k*-transformation* $(H(x), S(x)y + T(x))$.

3.2.2 The case of a single fixed point

We start with the simplest situation where a local solution may be extended to the entire manifold M in all smoothness classes.

Theorem 3.7. *Let $G(x,y) = (F(x), g(x,y))$ be a C^k-diffeomorphism of $\mathbb{R}^2$ and* $\operatorname{Fix} F = \{x_0\}$. *If equation* (3.0.1) *has a local C^k-solution at x_0, then it has a C^k-solution on the entire real line.*

We omit the proof, since it is similar to the proofs of the above theorems. Combining Theorem 3.7 with theorems on local solvability from Section 3.1 we obtain

Corollary 3.4. *Let $G(x,y) = (F(x), g(x,y))$ be a C^k-diffeomorphism of $\mathbb{R}^2$,* $\operatorname{Fix} F = \{x_0\}$, *and let, depending on k, conditions of either Theorem* 3.2, *or Theorem* 3.3, *or Theorem* 3.4 *hold. Then equation* (3.0.1) *has a C^k-solution on the entire real line.*

For linear equations with a unique hyperbolic fixed point there exists a classification similar to the previous section.

Corollary 3.5. *Let $G(x,y) = (F(x), a(x)y + \gamma(x)))$, let x_0 be a unique hyperbolic fixed point for a C^k-diffeomorphism F and let F be locally C^k-conjugate at x_0 with its linear part. Assume that a is a non-vanishing C^k-function and γ is a C^k-function. Then there exists a linear C^k-transformation $\Phi(x,y) = (H(x),\ S(x)y)$ reducing* (3.2.4) *to the form*

$$\psi(\lambda x) = \mu\psi(x) + \tilde{\gamma}(x) \tag{3.2.7}$$

where $\lambda = F'(x_0)$, $\mu = a(x_0)$ and $\tilde{\gamma}$ is a C^k-function.

Proof. We can assume that $x_0 = 0$ and $F(x) = \lambda x$. Then $S(x)$ must satisfy the equation

$$S(\lambda x) = \frac{a(x)}{\mu} S(x).$$

For $\tau(x) = \ln S(x)$ we have

$$\tau(\lambda x) = \tau(x) + b(x), \quad b(x) = \ln \frac{a(x)}{\mu}.$$

Since $b(0) = 0$, the latter equation is formally solvable at $x_0 = 0$. By Corollary 3.4 it has a C^k-solution τ on the entire real line. Using the transformation $S(x,y) = (x, y\exp(\tau(x))$ we reduce (3.2.4) to the form

$$\psi(\lambda x) = \mu\psi(x) + \gamma(x)\exp(-\tau(\lambda x)).$$

□

Let us now consider equation (3.0.1) on the unit circle. As above we will assume that $F : \mathbb{T} \to \mathbb{T}$ is a C^k-diffeomorphism with a unique fixed point x_0 and will show that the situation is different comparing with the real line. It is evident that x_0 is a point of overflow and $F'(x_0) = 1$. For every neighborhood $U \neq \mathbb{T}$ of x_0 we denote by U_- and U_+ its open semi-neighborhoods according to the standard orientation of $\mathbb{T}$.

Assume that φ_0 is a local C^k-solution of (3.0.1) in U. If the mapping $G(x, y) = (F(x), g(x, y))$ is a C^k-diffeomorphism of $\mathbb{T} \times \mathbb{R}$, then we can continue φ_0 from U_+ as a C^k-solution φ_+ on $\mathbb{T}\backslash\{x_0\}$. Since φ_+ coincides with φ_0 in U_+, we have

$$\lim_{x\to x_0;\ x\in U_+} \varphi_+^{(j)}(x) = \varphi_0^{(j)}(x_0), \qquad j = 0, \ldots, k,$$

but, generally speaking, the functions $\varphi_+^{(j)}(x)$ have no one-side limits in U_- as $x \to x_0$. Similarly, for the continuation of φ_0 from U_- as a C^k–solution φ_- on $\mathbb{T}\backslash\{x_0\}$ we have

$$\lim_{x\to x_0;\ x\in U_-} \varphi_-^{(j)}(x) = \varphi_0^{(j)}(x_0), \qquad j = 0, \ldots, k.$$

If the difference $\varphi_+ - \varphi_-$ is identically equal to zero on $\mathbb{T}\backslash\{x_0\}$, then φ_0 is continued as a global solution of (3.0.1). Generally speaking, it does not vanish, which is an obstacle to global solvability. We will give an explicit description of such obstacles for linear equations

$$\varphi(F(x)) - a(x)\varphi(x) = \gamma(x), \qquad x \in \mathbb{T}, \tag{3.2.8}$$

with a non-vanishing coefficient a.

For every local C^k-solution φ_0 of equation (3.2.8) at x_0 we set

$$\rho_\gamma(x; \varphi_0) = \varphi_+(x) - \varphi_-(x), \qquad x \in \mathbb{T}\backslash\{x_0\},$$

and note that the function $\rho_\gamma(x; \varphi_0)$ is a solution of the homogeneous equation

$$\psi(F(x)) - a(x)\psi(x) = 0 \tag{3.2.9}$$

in $\mathbb{T}\backslash\{x_0\}$. If $\tilde{\varphi}_0$ is another local C^k–solution of (3.2.8) with the same function γ and if $\tilde{\varphi}_+$ and $\tilde{\varphi}_-$ are its respective continuations, then

$$\rho_\gamma(x; \varphi_0) - \rho_\gamma(x; \tilde{\varphi}_0) = h_+(x) - h_-(x), \qquad x \in \mathbb{T}\backslash\{x_0\},$$

where the functions $h_\pm = \varphi_\pm - \tilde{\varphi}_\pm$ satisfy equation (3.2.9) in $\mathbb{T}\backslash\{x_0\}$ and coincide with the difference $\varphi_0 - \tilde{\varphi}_0$ in $U_\pm$.

Denote by $C^k_{x_0}(\mathbb{T})$ the set of all pairs $h_\pm \in C^k(\mathbb{T}\backslash\{x_0\})$ satisfying the homogeneous equation (3.2.9) and coinciding with a function from $C^k(U)$ in respective semi-neighborhoods $U_\pm$. In a non-analytic case the latter just means that for every $j = 0, 1, \ldots, k$ the finite limits

$$\lim_{x\to x_0;\ x\in U_+} h_+^{(j)}(x), \qquad \lim_{x\to x_0;\ x\in U_-} h_-^{(j)}(x)$$

exist and coincide. In any case, the function

$$\psi(x) = \begin{cases} h_+(x), & x \in U_+, \\ \lim\limits_{x \to x_0;\, x \in U_+} h_+(x) = \lim\limits_{x \to x_0;\, x \in U_-} h_-(x), & x = x_0, \\ h_-(x), & x \in U_-, \end{cases}$$

belongs to $C^k(U)$ for every pair $h_\pm \in C^k_{x_0}(\mathbb{T})$.

We will say that solutions ρ, $\tilde{\rho} \in C^k(\mathbb{T}\backslash\{x_0\})$ of (3.2.9) are *equivalent* if there exists a pair $h_\pm \in C^k_{x_0}(\mathbb{T})$ such that

$$\rho(x) - \tilde{\rho}(x) = h_+(x) - h_-(x), \qquad x \in \mathbb{T}\backslash\{x_0\}.$$

As we have just seen the functions $\rho_\gamma(x; \varphi_0)$ and $\rho_\gamma(x; \tilde{\varphi}_0)$ corresponding to local solutions φ_0 and $\tilde{\varphi}_0$ of equation (3.2.8) at x_0 with the same function γ are equivalent.

Example 3.9. Let us find the function $\rho_\gamma(x; \varphi_0)$ for a local solution φ_0 at x_0 of the cohomological equation

$$\varphi(F(x)) - \varphi(x) = \gamma(x), \qquad x \in \mathbb{T}. \tag{3.2.10}$$

To be definite, let x_0 attract points of U_- and hence repel points of U_+. Then the equality

$$\varphi_0(x) = \varphi_0(F^n(x)) - \sum_{j=0}^{n-1} \gamma(F^j(x))$$

holds for every $x \in U_-$ and $n = 1, 2, \ldots$ Since $\varphi_0 \in C^k(U)$, the sum here converges to a C^k-function in U_- as $n \to \infty$, and we obtain a representation

$$\varphi_0(x) = \varphi_0(x_0) - \sum_{j=0}^{\infty} \gamma(F^j(x)), \quad x \in U_-.$$

As a matter of fact, the series in the right-hand side of this identity is defined for all $x \in \mathbb{T}$ and the function

$$\varphi_-(x) = \varphi_0(x_0) - \sum_{j=0}^{\infty} \gamma(F^j(x)), \quad x \in \mathbb{T}\backslash\{x_0\},$$

is the continuation of φ_0 from U_-. Similarly, the function

$$\varphi_+(x) = \varphi_0(x_0) + \sum_{j=1}^{\infty} \gamma(F^{-j}(x)), \quad x \in \mathbb{T}\backslash\{x_0\},$$

is the continuation of φ_0 from U_+. We conclude that if equation (3.2.10) has a local C^k-solution at x_0, then the series

$$\sum_{j=-\infty}^{\infty} \gamma(F^j(x))$$

converges for all $x \in \mathbb{T}$, its sum belongs to the space $C^k(\mathbb{T}\backslash\{x_0\})$ and vanishes at x_0. Hence

$$\rho_\gamma(x;\varphi_0) = \sum_{j=-\infty}^{\infty} \gamma(F^j(x))$$

and the function ρ_γ does not depend on a choice of a local solution φ_0. Note that, generally speaking, the latter series is not continuous at x_0.

Let now $h_\pm \in C^k_{x_0}(\mathbb{T})$. Then

$$h_\pm(x) = h_\pm(F^n(x)), \qquad x \in \mathbb{T}\backslash\{x_0\},$$

and since x_0 attracts points of U_-,

$$h_\pm(x) = \lim_{n\to\mp\infty} h_\pm(F^n(x)) = \lim_{x\to x_0;\, x\in U_\pm} h_\pm(x).$$

Therefore h_+ and h_- are constant functions and functions ρ and $\tilde{\rho}$ are equivalent if and only if their difference is constant. In particular, $\rho_\gamma(x;\varphi_0)$ is equivalent to the zero function if and only if

$$\sum_{j=-\infty}^{\infty} \gamma(F^j(x)) = \text{const}, \qquad x \in \mathbb{T}\backslash\{x_0\}. \tag{3.2.11}$$

Theorem 3.8. *Let $F : \mathbb{T} \to \mathbb{T}$ be a C^k-diffeomorphism with a unique fixed point x_0 and let $\gamma \in C^k(\mathbb{T})$ be such that equation (3.2.8) is locally C^k-solvable at x_0. Then it has a global C^k-solution if and only if for every one of its local C^k-solutions φ_0 at x_0, the function $\rho_\gamma(x;\varphi_0)$ is equivalent to the zero function.*

For every function $\rho \in C^k(\mathbb{T}\backslash\{x_0\})$ satisfying the homogeneous equation (3.2.9) there exist a function $\gamma \in C^k(\mathbb{T})$ and a local solution φ_0 of (3.2.8) at x_0 with $\rho(x) = \rho_\gamma(x;\varphi_0)$.

Proof. Let $h_\pm \in C^k_{x_0}(\mathbb{T})$ be such that

$$\rho_\gamma(x;\varphi_0) = h_+(x) - h_-(x), \qquad x \in \mathbb{T}\backslash\{x_0\}.$$

If, as above, φ_+ and φ_- are the respective continuations of φ_0 on $\mathbb{T}\backslash\{x_0\}$, then the functions $\varphi_+ - h_+$ and $\varphi_- - h_-$ belong to the space $C^k(\mathbb{T}\backslash\{x_0\})$, satisfy equation (3.2.8) and coincide on $\mathbb{T}\backslash\{x_0\}$. Since

$$\lim_{x\to x_0;\, x\in U_+} (\varphi_+^{(j)}(x) - h_+^{(j)}(x)) = \varphi_0^{(j)}(x_0) - \lim_{x\to x_0;\, x\in U_+} h_+^{(j)}(x)$$

$$= \varphi_0^{(j)}(x_0) - \lim_{x\to x_0;\, x\in U_-} h_-^{(j)}(x) = \lim_{x\to x_0;\, x\in U_-} (\varphi_-^{(j)}(x) - h_-^{(j)}(x))$$

for $j = 0, \ldots, k$, these functions define a global C^k-solution of (3.2.8).

Conversely, if φ is a global solution of (3.2.8), then $\varphi_+ = \varphi_- = \varphi$ and hence $\rho_\gamma(x;\varphi) \equiv 0$.

Let now a function $\rho \in C^k(\mathbb{T}\backslash\{x_0\})$ satisfy the homogeneous equation (3.2.9). We fix an arbitrary neighborhood $U \neq \mathbb{T}$ of x_0 and set

$$\tau(x) = \begin{cases} -\rho(x), & x \in U_+, \\ 0, & x \in U_-. \end{cases}$$

The function τ belongs to the space $C^k(U\backslash\{x_0\})$ and may be represented in the form

$$\tau(x) = \psi(x) - \varphi_0(x), \qquad x \in U\backslash\{x_0\},$$

where $\psi \in C^k(\mathbb{T}\backslash\{x_0\})$ and $\varphi_0 \in C^k(U)$: for functions from classes C^k with $k < \infty$ and $k = \infty$, such decompositions may be obtained using a partition of identity; for analytic functions this is a version of the Cousin problem [46].

Let us define

$$\gamma(x) = \begin{cases} \varphi_0(F(x)) - a(x)\varphi_0(x), & x \in U, \\ \psi(F(x)) - a(x)\psi(x), & x \in \mathbb{T}\backslash\{x_0\}. \end{cases}$$

Since τ satisfies equation (3.2.9) in $U\backslash\{x_0\}$, γ is a well-defined C^k-function on $\mathbb{T}$ and φ_0 is a local solution of (3.2.8) at x_0. To find $\rho_\gamma(x;\varphi_0)$ we note that

$$\psi(x) - \varphi_+(x) = \psi(x) - \varphi_0(x) = -\rho(x), \qquad x \in U_+,$$

and

$$\psi(x) - \varphi_-(x) = \psi(x) - \varphi_0(x) = 0, \qquad x \in U_-.$$

Since the functions $\psi - \varphi_+$ and $-\rho$ are solutions of the homogeneous equation on $\mathbb{T}\backslash\{x_0\}$ coinciding on U_+, they coincide everywhere on $\mathbb{T}\backslash\{x_0\}$. Similarly, $\psi(x) - \varphi_-(x) = 0$ for all $x \in \mathbb{T}\backslash\{x_0\}$. Therefore $\rho_\gamma(x;\varphi_0) = \varphi_+(x) - \varphi_-(x) = \rho(x)$ which completes the proof of the theorem. □

Combining the description of $C^k_{x_0}(\mathbb{T})$ given in Example 3.9 with Theorem 3.8 we arrive at the following proposition.

Corollary 3.6. *Let $F : \mathbb{T} \to \mathbb{T}$ be a C^k-diffeomorphism with a unique fixed point x_0. Given a function $\gamma \in C^k(\mathbb{T})$, the cohomological equation* (3.2.10) *has a global C^k-solution if and only if it is locally C^k-solvable at x_0 and* (3.2.11) *holds.*

This statement implies that if equation (3.2.10) is locally C^k-solvable and has a global continuous solution, then it is globally C^k-solvable as well.

It follows from the second part of Theorem 3.8 that there exists an infinite-dimensional manifold of C^k-functions γ for which, on one hand, there exists a local C^k-solution φ_0 of (3.2.10), but on the other hand, $\rho_\gamma(x,\varphi_0)$ is not a constant function. For such γ's no continuous global solution does exist.

The following theorem contains conditions which guarantee the global solvability of (3.2.8) with an arbitrary function γ.

Theorem 3.9. *Let either $k < \infty$ or $k = \infty$, let $F : \mathbb{T} \to \mathbb{T}$ be a C^k-diffeomorphism with* $\operatorname{Fix} F = \{x_0\}$ *and let a be a non-vanishing C^k-function with $|a(x_0)| \neq 1$. Then equation* (3.2.8) *has a unique global C^k-solution for every function $\gamma \in C^k(\mathbb{T})$.*

Proof. Since $F'(x_0) = 1$ and $|a(x_0)| \neq 1$, it follows from Theorems 3.2 and 3.3 that for every $\gamma \in C^k(\mathbb{T})$ there exists a local C^k-solution of equation (3.2.8). Moreover, if, to be definite, $|a(x_0)| < 1$, then there exist $N > 0$ and $\mu \in (0,1)$ such that

$$|a(x)a(F^{-1}(x)) \dots a(F^{-n}(x))| \leq \mu^n, \qquad n \geq N.$$

Therefore the series

$$\varphi_0(x) = \gamma(F^{-1}(x)) + \sum_{n=1}^{\infty} a(F^{-1}(x)) \dots a(F^{-n}(x))\gamma(F^{-n-1}(x))$$

converges uniformly on $\mathbb{T}$ and is a unique continuous global solution of (3.2.8). In order to complete the proof of Theorem 3.9 it remains to check that $\varphi_0 \in C^k(\mathbb{T})$. To this end we could use the same arguments as in the proof of Theorems 3.2 and 3.3. Nevertheless we prefer an approach based on the previous theorem. According to this theorem it is sufficient to prove that every C^k-solution ρ of the homogeneous equation (3.2.9) on $\mathbb{T}\backslash\{x_0\}$ is equivalent to the zero function.

Let us, to be definite, assume that x_0 repels points from U_+ and prove that

$$\lim_{x \to x_0;\, x \in U_-} \rho^{(j)}(x) = 0, \qquad j = 0, \dots, k. \tag{3.2.12}$$

Because of the local character of the problem it is sufficient to consider instead of (3.2.9) the homogeneous equation

$$h(t) = b(t)h(G(t)), \qquad t \neq 0,$$

in a punctured neighborhood U of $t = 0$ with a local C^k-diffeomorphism

$$G : \mathbb{R} \to \mathbb{R};\ G(0) = 0;\ G'(0) = 1;\ G^{-n}(t) \to 0,\ t > 0,\ n \to +\infty,$$

and a local C^k-function b satisfying $|b(t)| \leq \mu < 1$, and to prove that every of its C^k-solutions h in U satisfies conditions

$$\lim_{t \to 0+} h^{(j)}(t) = 0, \qquad j = 0, \dots, k. \tag{3.2.13}$$

For a proof we fix $y > 0$ such that $[y, G(y)] \subset U$ and for every $t \in (0, y)$ denote by $n(t)$ the smallest integer $n = n(t)$ satisfying $G^n(t) \in [y, G(y)]$. Since G is a monotonic and fixed-point free diffeomorphism in $(0, y]$, the function $n(t)$ monotonically increases and tends to $+\infty$ as $t \to 0+$. The iterated identity

$$h(t) = b(t) \dots b(G^{n-1}(t))h(G^n(t)), \qquad n = n(t),$$

implies

$$|h(t)| \leq c\mu^n, \quad t \in [y, G(y)], \quad c = \max_{s \in [y, G(y)]} |h(s)|.$$

Given an arbitrary $\epsilon > 0$, we first find N such that $c\mu^N \leq \epsilon$, then $\delta \in (0, y)$ satisfying $n(\delta) \geq N$ and obtain the estimate

$$|h(t)| \leq c\mu^{n(t)} \leq c\mu^N \leq \epsilon, \quad t \in (0, \delta],$$

proving (3.2.13) for $j = 0$.

If relations (3.2.13) are already proven for $j = 0, 1, \ldots, m-1$, $m \leq k$, then

$$h^{(m)}(t) = b_m(t)h^{(m)}(G(t)) + \gamma_m(t)$$

where

$$b_m(t) = b(t)(G'(t))^m, \qquad \lim_{t\to 0+} \gamma_m(t) = 0.$$

Since $G'(0) = 1$ we may choose a neighborhood $U_m \subset U$ of $t = 0$ such that

$$|b_m(t)| \leq \mu_m < 1, \qquad t \in U_m.$$

We fix $y > 0$ such that $[y, G(y)] \subset U_m$ and define $n(t)$ as above. Starting from an arbitrary $\epsilon > 0$ we choose $\delta \in (0, y)$ such that $|\gamma_m(t)| \leq \epsilon$ for $t \in [0, \delta]$. For every $t \in (0, \delta]$ denote by $l(t)$ the largest integer such that $G^p(t) \leq \delta$, $p = 0, 1, \ldots, l(t)$. It is evident that $l(t) \leq n(t)$ and $l(t) \to +\infty$. Now the iterated equation yields the estimate

$$|h^{(m)}(t)| \leq \mu_m^{n(t)} c_m + \sum_{p=0}^{n(t)-1} \mu_m^p |\gamma_m(G^p(t))|$$

$$\leq \mu_m^{n(t)} c_m + \epsilon \sum_{p=0}^{l(t)} \mu_m^p + M \sum_{p=l(t)+1}^{n(t)-1} \mu_m^p, \quad t \in (0, \delta],$$

where

$$c_m = \max_{s\in[y,G(y)]} |h^{(m)}(s)|, \qquad M = \max_{s\in[0,G(y)]} |\gamma_m(s)|.$$

Therefore

$$|h^{(m)}(t)| \leq \mu_m^{n(t)} c_m + \frac{\epsilon + M\mu_m^{l(t)}}{1-\mu_m}, \quad t \in (0, \delta],$$

and

$$\limsup_{t\to 0+} |h^{(m)}(t)| \leq \frac{\epsilon}{1-\mu_m}$$

completing the proof of (3.2.13) and hence of (3.2.12). Therefore for a solution ρ of equation (3.2.9) on $\mathbb{T}\backslash x_0$, the functions $h_+(x) \equiv 0$ and $h_-(x) = -\rho(x)$ form a pair from the set $C^k_{x_0}(\mathbb{T})$. Since $\rho = h_+ - h_-$, we conclude that ρ is equivalent to the zero function. □

Let us note that, as we have seen in Example 3.6, in the analytic case the condition $|a(x_0)| \neq 1$ does not guarantee the local solvability of equation (3.2.8).

3.3 Gluing method for linear equations with several fixed points

The methods described in Sections 2.4.3 and 2.4.4 for a classification of diffeomorphisms may be used for a study of equation (3.0.1) with several fixed points. As above, we restrict ourselves to linear equations only.

Given a C^k-mapping $F : M \to M$ and a C^k-function $a : M \to \mathbb{R}$, denote

$$L\varphi(x) = \varphi(F(x)) - a(x)\varphi(x), \quad x \in M.$$

The linear operator L is well defined in the space $C^k(M)$ of all C^k-functions $\varphi : M \to \mathbb{R}$. Our aim is to state conditions of the solvability of the equation

$$L\varphi = \gamma. \tag{3.3.1}$$

Let $\{U_j\}$ be an open F-invariant covering of M, either finite or countable. We restrict equation (3.3.1) to U_i and denote by $C(k, \{U_i\}, L)$ the space of all C^k-functions $\gamma : M \to \mathbb{R}$ such that every restricted equation has a C^k-solution $\varphi_i : U_i \to \mathbb{R}$. Then the differences

$$r_{ij}(x) = \varphi_j(x) - \varphi_i(x), \quad x \in U_i \bigcap U_j, \tag{3.3.2}$$

are C^k-functions defined on the intersections $U_i \bigcap U_j$ and satisfying the homogeneous equations

$$Lr_{ij}(x) = 0, \quad x \in U_i \bigcap U_j. \tag{3.3.3}$$

In addition, the identities

$$r_{ij}(x) = -r_{ji}(x), \quad r_{ij}(x) + r_{jk}(x) + r_{ki}(x) = 0, \quad x \in U_i \bigcap U_j \bigcap U_k,$$

are valid. A collection $r = \{ r_{ij}(x) \}$ of solutions to equation (3.3.3) satisfying the latter relations is called a *cocycle*. We denote by $Z(k, \{U_i\}, L)$ the space of all cocycles corresponding to the covering $\{U_i\}$ and the operator L.

Two cocycles $\tilde{r} = \{\tilde{r}_{ij}\}$ and $r = \{r_{ij}\}$ are said to be *equivalent* if there exists a collection of C^k-functions $\{\psi_i(x)\}$ satisfying the homogeneous equation

$$L\psi(x) = 0, \quad x \in U_i. \tag{3.3.4}$$

such that

$$\tilde{r}_{ij}(x) - r_{ij}(x) = \psi_j(x) - \psi_i(x), \quad x \in U_i \bigcap U_j. \tag{3.3.5}$$

Denote by $H^1(k, \{U_i\}, L)$ the quotient space

$$Z(k, \{U_i\}, L) \,/\, B(k, \{U_i\}, L)$$

where $B(k, \{U_i\}, L)$ is the space of all cocycles equivalent to zero.

If $\gamma \in C(k, \{U_i\}, L)$ and $\{\varphi_i\}$ is a collection of local solutions to equation (3.3.1) on the covering $\{U_i\}$, then it follows from (3.3.2) and (3.3.3), that the collection $\{\varphi_j - \varphi_i\}$ is a cocycle. If $\{\tilde{\varphi}_i\}$ is another collection of local solutions corresponding to the same function γ, then $\psi_i = \varphi_i - \tilde{\varphi}_i$ satisfies the homogeneous equation (3.3.4). Therefore cocycles $\{\varphi_j - \varphi_i\}$ and $\{\tilde{\varphi}_j - \tilde{\varphi}_i\}$ are equivalent and hence to every function $\gamma \in C(k, \{U_i\}, L)$ there corresponds an equivalence class $r[\gamma] \in H^1(k, \{U_i\}, L)$.

Theorem 3.10. *Equation* (3.3.1) *with* $\gamma \in C(k, \{U_i\}, L)$ *has a global* C^k*-solution if and only if* $r[\gamma] = 0$.
The mapping

$$r : C(k, \{U_i\}, L) \to H^1(k, \{U_i\}, L) \tag{3.3.6}$$

is surjective.

Proof. If φ is a global C^k-solution of equation (3.3.1) and φ_i is a local C^k-solution of its restriction to U_i, then the difference $\psi_i = \varphi - \varphi_i$ satisfies the homogeneous equation (3.3.4). Since $\varphi_j - \varphi_i = \psi_j - \psi_i$, we obtain $r[\gamma] = 0$.

Conversely, let $r[\gamma] = 0$. Then for every local C^k-solution φ_i of equation (3.3.1) on U_i we have

$$\varphi_j(x) - \varphi_i(x) = \psi_j(x) - \psi_i(x), \quad x \in U_i \bigcap U_j,$$

where the ψ_i satisfy (3.3.4) and hence the function

$$\varphi(x) = \varphi_i(x) - \psi_i(x), \quad x \in U_i,$$

is a well-defined C^k-solution of (3.3.1).

Let now an equivalence class $c \in H^1(k, \{U_i\}, L)$ be given and let a cocycle $\{r_{ij}(x)\}$ be a representative of c. In order to prove the second statement of Theorem 3.10 it is sufficient to find C^k-functions $\varphi_i(x)$, $x \in U_i$, such that

$$r_{ij}(x) = \varphi_j(x) - \varphi_i(x), \quad x \in U_i \bigcap U_j. \tag{3.3.7}$$

Equations (3.3.7) compose a one-dimensional version of the Cousin Problem. It is well known (cf., [44], [46], [71]) that it has solutions in classes of smooth and analytic functions. If $\{\varphi_i\}$ are such solutions, then

$$\gamma(x) = L\varphi_i(x), \quad x \in U_i,$$

is a well-defined C^k-function belonging to $C(k, \{U_i\}, L)$, and $r[\gamma] = c$. □

If the set of non-wandering points of a mapping F consists of isolated fixed points only, then the description of the cocycle space may be simplified.

Let $F : M \to M$ be an orientation-preserving C^k-diffeomorphism and let the set

$$\operatorname{Fix} F = \{x_1, \dots, x_n\}$$

be ordered according to the orientation on M. For $M = \mathbb{R}$ we set

$$U_1 = (-\infty, x_2), \quad U_n = (x_{n-1}, +\infty),$$

while for $M = \mathbb{T}$, we define $x_{n+1} = x_1$ and

$$U_1 = U_{n+1} = (x_n, x_2), \quad U_n = (x_{n-1}, x_1).$$

Besides, let

$$U_i = (x_{i-1}, x_{i+1}), \quad i = 2, \dots, n-1, \tag{3.3.8}$$

for both cases. Since $U_i \bigcap U_{i+1} = (x_i, x_{i+1})$, $i = 1, \dots, n$ and $U_i \bigcap U_j = \emptyset$ if $|i - j| \geq 2$, the cocycle space $Z(k, \{U_i\}, L)$ consists of collections $\{r_{ij}(x)\}$ such that the functions $\rho_i(x) = r_{i,i+1}(x)$ satisfy the homogeneous equation

$$\rho_i(F(x)) = a(x)\rho_i(x), \quad x \in U_i \bigcap U_{i+1},$$

and $r_{ij} = 0$ for $|i - j| \geq 2$. It means that if $M = \mathbb{R}$ then every cocycle from $Z(k, \{U_i\}, L)$ is representable as a collection $(\rho_1(x), \dots, \rho_{n-1}(x))$ and if $M = \mathbb{T}$ then it is of the form $(\rho_1(x), \dots, \rho_n(x))$.

Let us define $M_F = M \setminus \operatorname{Fix} F$ for $M = \mathbb{T}$ and $M_F = [x_1, x_n] \setminus \operatorname{Fix} F$ for $M = \mathbb{R}$. If we set

$$h(x) = \rho_i(x), \quad x \in (x_i, x_{i+1}), \quad i = 1, \dots, n-1,$$

then we obtain a function belonging to $C^k(M_F)$. Therefore the space $Z(k, \{U_i\}, L)$ may be identified with the space of functions $h \in C^k(M_F)$ satisfying the homogeneous equation

$$h(F(x)) - a(x)h(x) = 0, \quad x \in M_F. \tag{3.3.9}$$

Assume $a(x)$ does not vanish in (x_i, x_{i+1}) for some value of i. If $\Psi : (x_i, x_{i+1}) \to \mathbb{R}$ is a straightening C^k-diffeomorphism for F, i.e.,

$$\Psi F \Psi^{-1}(t) = t + 1, \qquad t \in \mathbb{R},$$

and if θ is a C^k-solution of the cohomological equation

$$\theta(t+1) - \theta(t) = \ln |a(\Psi^{-1}(t))|, \qquad t \in \mathbb{R},$$

then the function

$$h(x) = \begin{cases} \tau(\Psi(x)) \exp\, \theta(\Psi(x)), & x \in (x_i, x_{i+1}), \\ 0, & x \notin (x_i, x_{i+1}), \end{cases} \tag{3.3.10}$$

belongs to $Z(k, \{U_i\}, L)$ for every C^k-function τ which is 1-periodic if $a(x) > 0$ and 1-anti-periodic otherwise. Hence the space $Z(k, \{U_i\}, L)$ is infinite-dimensional if a does not vanish on (x_i, x_{i+1}) at least for one value of i. Moreover, if a does not vanish on M at all, then this space is spanned by functions (3.3.10).

A detailed description of the quotient space $H^1(k, \{U_i\}, L)$ depends on additional properties of F and a. In the following subsections we present such a description for the cohomological equation ($a(x) \equiv 1$) and for equation (3.2.4) with F having hyperbolic fixed points only.

3.3.1 Cohomological equation

If $F : M \to M$ is an orientation-preserving C^k-diffeomorphism with isolated fixed points $x_1, \dots, x_n$, then the homogeneous equation

$$\psi(F(x)) - \psi(x) = 0, \quad x \in U_i,$$

has constant solutions only. Hence two cocycles h and $\tilde{h}$ are equivalent if and only if

$$h(x) - \tilde{h}(x) = \text{const}, \qquad x \in (x_i, x_{i+1}),$$

and the space $B(k, \{U_i\}, L)$ may be identified with the space of all functions on M_F of the form

$$\rho(x) = c_i, \qquad x \in (x_i, x_{i+1}),$$

with $i = 1, \dots, n-1$ if $M = \mathbb{R}$ and $i = 1, \dots, n$ for $M = \mathbb{T}$. Since $\dim B(k, \{U_i\}, L) < \infty$, the space $H^1(k, \{U_i\}, L)$ is infinite-dimensional.

If now $\gamma \in C(k, \{U_i\}, L)$, then a solution of equation

$$\varphi(F(x)) - \varphi(x) = \gamma(x), \quad x \in U_i, \tag{3.3.11}$$

is determined up to a constant.

More exactly, we have

$$\varphi_i(x) = \varphi_i(F^k(x)) - \sum_{s=0}^{k-1} \gamma(F^s(x)), \quad x \in U_i,$$

and if x_i attracts then

$$\varphi_i(x) = -\sum_{s=0}^{\infty} \gamma(F^s(x)) + \varphi_i(0), \quad x \in U_i. \tag{3.3.12}$$

Similarly, if x_i repels then

$$\varphi_i(x) = \sum_{s=1}^{\infty} \gamma(F^{-s}(x)) + \varphi_i(0), \quad x \in U_i. \tag{3.3.13}$$

If x_i is a point of overflow attracting on the interval (x_{i-1}, x_i) and repelling on the interval (x_i, x_{i+1}), then

$$\varphi_i(x) = \varphi_i(0) + \begin{cases} -\sum\limits_{s=0}^{\infty} \gamma(F^s(x)), & x \in (x_{i-1}, x_i), \\ \sum\limits_{s=1}^{\infty} \gamma(F^{-s}(x)), & x \in (x_i, x_{i+1}). \end{cases}$$

In any case for every cocycle $\{\rho_i(x)\}$ corresponding to γ we have

$$\rho_i(x) = \pm \sum_{s=-\infty}^{\infty} \gamma(F^s(x)) + c_i, \quad x \in (x_i, x_{i+1}),$$

with some constants c_i. We conclude that for every $\gamma \in C(k, \{U_i\}, L)$ the series

$$\sum_{s=-\infty}^{\infty} \gamma(F^s(x)), \qquad x \in M_F,$$

converges to a C^k-function on M_F and represents a cocycle corresponding to γ. As a result we arrive at the following statement.

Theorem 3.11. *Let $F : M \to M$ be an orientation-preserving C^k-diffeomorphism with fixed points $x_1, \dots, x_n$. Then the cohomological equation*

$$\varphi(F(x)) - \varphi(x) = \gamma(x)$$

has a global C^k-solution on M if and only if it has a local C^k-solution at every point x_j and

$$h_\gamma(x) \equiv \sum_{s=-\infty}^{\infty} \gamma(F^s(x)) = \text{const}, \ x \in (x_i, x_{i+1}).$$

This statement implies that if for $\gamma \in C(k, \{U_i\}, L)$ there exists a global continuous solution, then there exists a global C^k-solution as well.

Since the space $H^1(k, \{U_i\}, L)$ is infinite-dimensional, there exists an infinite-dimensional variety of functions $\gamma \in C(k, \{U_i\}, L)$ such that corresponding cohomological equations have no global continuous solutions. To construct an explicit example of such a function on $\mathbb{R}$, assume that the point x_1 attracts from the right and hence x_2 repels from the left. Fix a point $z \in (x_1, x_2)$. Then

$$x_1 < F(z) < z < F^{-1}(z) < x_2.$$

Furthermore, let $\theta \in (z, F^{-1}(z))$ and $\varepsilon > 0$ be such that $[\theta - \varepsilon, \theta + \varepsilon] \subset (z, F^{-1}(z))$. Choose a C^∞-function $\gamma : \mathbb{R} \to [0, 1]$ such that

$$\gamma(x) = \begin{cases} 1, & x \in [\theta - \varepsilon, \theta + \varepsilon], \\ 0, & x \notin [z, F^{-1}(z)]. \end{cases}$$

Since $\gamma(x) = 0$ in a neighborhood of every fixed point, the cohomological equation

$$\varphi(F(x)) - \varphi(x) = \gamma(x)$$

has a C^∞-solution on U_i for every $i = 1, \ldots, n$, but it has no global continuous solution. Indeed,

$$h_\gamma(\theta) = 1, \quad h_\gamma(z) = 0,$$

and $h_\gamma(x)$ is not constant on (x_1, x_2).

3.3.2 Equations with hyperbolic fixed points

In this section we consider equation (3.2.4) where F is an orientation-preserving C^k-diffeomorphism with $k \geq 1$, hyperbolic fixed points $x_1, \ldots, x_n$ and a is a C^k-function non-vanishing on M. According to Theorem 2.5 if $k \geq 2$, then the restriction of F to U_j is C^k-conjugate with its linear part $F_j(x) = \lambda_j x$, $\lambda_j = F'(x_j)$. Our additional assumption is that such conjugacy holds for $k = 1$ as well. This assumption implies that for every j the restriction of equation (3.2.4) to U_j is C^k-equivalent to (3.2.7), which means that there exists an invertible transformation

$$\Phi(x, y) = (\Psi_j(x), S_j(x)y), \qquad x \in U_j, \tag{3.3.14}$$

with a C^k-diffeomorphism

$$\Psi_j : U_j \to \mathbb{R}, \quad \Psi_j(x_j) = 0, \tag{3.3.15}$$

and a C^k-function

$$S_j : U_j \to \mathbb{R}, \quad S_j(x_j) = 1, \quad S_j(x) \neq 0, \quad x \in U_j, \tag{3.3.16}$$

such that

$$\Psi_j F \Psi_j^{-1}(t) = \lambda_j t, \quad S_j(\Psi_j^{-1}(t)) = \frac{a(\Psi_j^{-1}(t))}{\mu_j} S_j(\Psi_j^{-1}(\lambda_j t)), \quad t \in \mathbb{R}, \tag{3.3.17}$$

where $\mu_j = a(x_j)$.

In this sense we say that *equation* (3.2.4) *is C^k-glued from equations* (3.2.7). Everyone of its solutions φ on U_j may be represented in the form

$$\varphi(x) = \frac{\tilde{\varphi}(\Psi_j(x))}{S_j(x)}, \qquad x \in U_j, \tag{3.3.18}$$

where $\tilde{\varphi}$ is a solution of (3.2.7) with

$$\tilde{\gamma}(t) = S_j(\Psi_j^{-1}(\lambda_j t))\gamma(\Psi_j^{-1}(t)), \qquad t \in \mathbb{R}. \tag{3.3.19}$$

In contrast to the cohomological equation treated in the previous section, the space $B(k, \{U_i\}, L)$ generated by equation (3.2.4) may be infinite-dimensional. As a result the space $H^1(k, \{U_i\}, L)$ may be finite-dimensional and even trivial. To study it we start with a description of C^k-solutions of the homogeneous equation (3.2.4) on U_j.

Let

$$\alpha_j = \frac{\ln|\mu_j|}{\ln\lambda_j}, \quad j = 1, \dots, n.$$

Then the general solution of the equation

$$\psi(\lambda_j t) = \mu_j \psi(t), \qquad t \in \mathbb{R}, \tag{3.3.20}$$

has the form

$$\psi(t) = \begin{cases} t^{\alpha_j}\cos(\theta_j \ln t)\tau_+\left(\dfrac{\ln t}{\ln\lambda_j}\right), & t > 0, \\ |t|^{\alpha_j}\cos(\theta_j \ln|t|)\tau_-\left(\dfrac{\ln|t|}{\ln\lambda_j}\right), & t < 0. \end{cases}$$

Here either $\theta_j = 0$ or $\theta_j = \pi(\ln\lambda_j)^{-1}$, depending on the sign of $a(x)$, and $\tau_\pm$ are arbitrary 1-periodic functions. It is easy to see that these formulas define a C^k-solution of equation (3.3.20) on U_j if and only if either $k < \alpha_j$ or $\alpha_j = s$ is a natural number, $\theta_j = 0$ and $\tau_\pm$ are constant functions; in the latter case $\psi(t) = ct^{\alpha_j}$. Hence a general solution of equation

$$\varphi(F(x)) - a(x)\varphi(x) = 0, \quad x \in U_j, \tag{3.3.21}$$

has the form

$$\varphi(x) = S_j^{-1}(x)\omega(x), \quad x \in U_j,$$

where

$$\omega(x) = \begin{cases} \Psi_j^{\alpha_j}(x)\cos(\theta_j \ln\Psi_j(x))\tau_+\left(\dfrac{\ln\Psi_j(x))}{\ln\lambda_j}\right), & x \in (x_j, x_{j+1}), \\ |\Psi_j(x)|^{\alpha_j}\cos(\theta_j \ln|\Psi_j(x)|)\tau_-\left(\dfrac{\ln|\Psi_j(x)|}{\ln\lambda_j}\right), & x \in (x_{j-1}, x_j). \end{cases}$$

Such a solution is a C^k-function on U_j if and only if either $k < \alpha_j$ or α_j is a natural number, $\theta = 0$, $\tau_+(t) \equiv \tau_-(t) \equiv c$, $c \in \mathbb{R}$. In the latter case a resonance relation $\mu_j = \lambda_j^s$ with $s = \alpha_j$ holds and all C^k-solutions are of the form

$$\varphi(x) = cS_j^{-1}(x)\Psi_j^{\alpha_j}(x), \quad c \in \mathbb{R}.$$

Of course, if a resonance exponent s does exist, then it is unique and the space $B(k, \{U_i\}, L)$ is finite-dimensional if and only if $k \geq \max\alpha_j, j = 1, \dots, n$.

To every cocycle $h \in Z(k, \{U_j\}, L)$, we relate the *modified cocycle*

$$\hat{h}(x) = \begin{cases} h(x), & x \in (x_j, x_{j+1}) \text{ if } k \geq \max(\alpha_j, \alpha_{j+1}), \\ 0, & x \in (x_j, x_{j+1}) \text{ if } k < \max(\alpha_j, \alpha_{j+1}), \end{cases}$$

where $1 \leq j \leq n-1$ if $M = \mathbb{R}$ and $1 \leq j \leq n$, $x_{n+1} = x_1$ if $M = \mathbb{T}$.

Theorem 3.12. *Let a C^k-diffeomorphism $F : M \to M$, $k \geq 1$, be such that all fixed points $x_1, \dots, x_n$ of F are hyperbolic, let all restrictions $F\big|_{U_j}$ be C^k-conjugate with their linear parts at x_j and let $a(x) \neq 0$, $x \in M$. Then every cocycle $h \in Z(k, \{U_j\}, L)$ is equivalent to its modified cocycle $\hat{h}$ and the space $\hat{B}(k, \{U_j\}, L) \subset B(k, \{U_j\}, L)$ of all modified cocycles equivalent to zero is finite dimensional.*

Proof. It is sufficient to check that if $k < \max(\alpha_j, \alpha_{j+1})$ then every cocycle of the form

$$(\underbrace{0, \dots, 0}_{j-1}, \rho(x), 0, \dots, 0) \tag{3.3.22}$$

is equivalent to zero.

Since $\rho(x)$ is a C^k-solution of the homogeneous equation on the interval (x_j, x_{j+1}), it may be represented either in the form

$$\rho(x) = S_j^{-1}(x)\, \Psi_j^{\alpha_j}(x) \cos(\theta \ln \Psi_j(x)) \tau_+ \left(\frac{\ln \Psi_j(x)}{\ln \lambda_j} \right),$$

if $k < \alpha_j$, or in the form

$$\rho(x) = S_{j+1}^{-1}(x)\, |\Psi_{j+1}(x)|^{\alpha_{j+1}} \cos(\theta \ln |\Psi_{j+1}(x)|) \tau_- \left(\frac{\ln |\Psi_{j+1}(x)|}{\lambda_{j+1}} \right)$$

if $k < \alpha_{j+1}$, with some 1-periodic C^k-functions $\tau_\pm$.

In the former case the function

$$\rho_1(x) = \begin{cases} \rho(x), & x \in (x_j, x_{j+1}), \\ 0, & x \in (x_{j-1}, x_j], \end{cases}$$

is a C^k-solution of (3.3.21) on U_j, while in the second case the function

$$\rho_2(x) = \begin{cases} \rho(x), & x \in (x_j, x_{j+1}), \\ 0, & x \in [x_{j+1}, x_{j+2}), \end{cases}$$

is a C^k-solution of (3.3.21) on U_{j+1}.

Now we set

$$\varphi_j(x) = -\rho_1(x),\ x \in U_j;\ \varphi_i(x) = 0,\ x \in U_i,\ i \neq j,$$

for $k < \alpha_j$ and

$$\varphi_{j+1}(x) = \rho_2(x),\ x \in U_{j+1};\ \varphi_j(x) = 0,\ x \in U_i, i \neq j+1,$$

for $k < \alpha_{j+1}$. According to this definition

$$\varphi_{i+1}(x) - \varphi_i(x) = 0, \quad x \in U_i \bigcap U_{i+1},\ i \neq j,$$

and

$$\varphi_{j+1}(x) - \varphi_j(x) = \rho(x), \quad x \in (x_j, x_{j+1}).$$

Therefore the cocycle (3.3.22) is equivalent to zero.

To prove that the space $\hat{B}(k, \{U_j\}, L)$ is finite-dimensional let us construct a system of cocycles spanning it. If for a given index j, $j = 1, \dots, n$, there are no resonance exponents $s \leq k$, we set $g_j(x) = 0$, $x \in (x_{j-1}, x_{j+1})$. Otherwise we set

$$g_j(x) = S_j^{-1}(x)\Psi_j^s(x), \qquad x \in (x_{j-1}, x_{j+1}), \tag{3.3.23}$$

and consider the cocycles

$$h_{2j-1}(x) = \begin{cases} g_j(x), & x \in (x_{j-1}, x_j), \\ 0, & x \notin (x_{j-1}, x_j), \end{cases} \tag{3.3.24}$$

$$h_{2j}(x) = \begin{cases} g_j(x), & x \in (x_j, x_{j+1}), \\ 0, & x \notin (x_j, x_{j+1}), \end{cases} \tag{3.3.25}$$

with $h_1 = 0$ for $M = \mathbb{R}$.

Assume that a cocycle $\hat{h} = (\rho_1(x), \dots, \rho_n(x))$ belongs to the space $\hat{B}(k, \{U_j\}, L)$. Being a modified cocycle it may be represented as a sum of cocycles $(0, \dots, \rho_j(x), \dots, 0)$ with $k \geq \max(\alpha_j, \alpha_{j+1})$. Since $\hat{h}$ is equivalent to zero,

$$\rho_j(x) = \varphi_{j+1}(x) - \varphi_j(x), \quad x \in (x_j, x_{j+1}),$$

where

$$\varphi_j(x) = c_j g_j(x), \qquad \varphi_{j+1}(x) = c_{j+1} g_{j+1}(x)$$

with some constants c_j, c_{j+1}. Hence $\hat{h}$ belongs to the subspace spanned by $h_1, \dots, h_{2n}$ and $\dim \hat{B}(k, \{U_j\}, L) \leq 2n$. □

Corollary 3.7. *Let all assumptions of Theorem* 3.12 *hold. Given a function* $\gamma \in C(k, \{U_j\}, L)$, *equation* (3.2.4) *has a global* C^k*-solution if and only if for every cocycle* h *generated by* γ *the modified cocycle* $\hat{h}$ *belongs to the space* $\hat{B}(k, \{U_j\}, L)$.

Now we can formulate conditions under which the local solvability of (3.2.4) at each fixed point of F implies its global solvability on M.

Theorem 3.13. *Let all assumptions of Theorem* 3.12 *be fulfilled and let*

$$\lambda_j = F'(x_j), \quad \mu_j = a(x_j), \qquad j = 1, \dots, n. \tag{3.3.26}$$

Then equation (3.2.4) *has a solution* $\varphi \in C^k(M, \mathbb{R})$ *for every function* $\gamma \in C(k, \{U_j\}, L)$ *if and only if*

$$k < \max\left(\frac{\ln|\mu_j|}{\ln\lambda_j}, \frac{\ln|\mu_{j+1}|}{\ln\lambda_{j+1}}\right), \quad j = 1, \dots, n-1,$$

for $M = \mathbb{R}$ *and, in addition to the previous condition,*

$$k < \max\left(\frac{\ln|\mu_n|}{\ln\lambda_n}, \frac{\ln|\mu_1|}{\ln\lambda_1}\right)$$

for $M = \mathbb{T}$.

Proof. If the inequalities hold, then all modified cocycles are equal to zero and hence all cocycles are equivalent to zero. To prove the "only if" part we note that the space of all cocycles $h = (\underbrace{0, \dots, 0}_{j-1}, \rho(x), 0, \dots, 0)$ with

$$\rho(x) = S_j^{-1}(x)\Psi_j^{\alpha_j}(x)\cos(\theta\ln\Psi_j(x))\tau\left(\frac{\ln\Psi_j(x)}{\ln\lambda_j}\right), \quad x \in (x_j, x_{j+1}),$$

and arbitrary 1-periodic C^k-functions τ is infinite-dimensional. On the other hand, if $k \geq \max(\alpha_j, \alpha_{j+1})$ for some j, then every such cocycle is modified and all of them cannot belong to the finite-dimensional space $\hat{B}(k, \{U_j\}, L)$. Let h be a modified cocycle not belonging to $\hat{B}(k, \{U_j\}, L)$. According to Theorem 3.10 there exists a function $\gamma \in C(k, \{U_j\}, L)$ such that the equivalence class $r[\gamma]$ contains h. Hence $r[\gamma] \neq 0$ and equation (3.2.4) with such a function γ has no global C^k-solutions. □

Another consequence of the previous results is a criterion of the global solvability for every C^k-function γ.

Theorem 3.14. *Let all assumptions of Theorem* 3.12 *hold. Then equation* (3.2.4) *has a* C^k*-solution for every* C^k*-function* γ *if and only if the following conditions hold:*

i) For all $s = 0, 1, \dots, k$, *and all* $j = 1, \dots, n$,

$$\mu_j \neq \lambda_j^s, \quad |\mu_j| \neq \lambda_j^k,$$

where λ_j *and* μ_j *are defined by* (3.3.26).

ii) For $M = \mathbb{R}$,

$$k < \max\left(\frac{\ln|\mu_j|}{\ln\lambda_j}, \frac{\ln|\mu_{j+1}|}{\ln\lambda_{j+1}}\right), \quad j = 1, \dots, n-1,$$

and, in addition to the previous condition,

$$k < \max\left(\frac{\ln|\mu_n|}{\ln\lambda_n}, \frac{\ln|\mu_1|}{\ln\lambda_1}\right)$$

for $M = \mathbb{T}$.

Proof. According to Corollary 3.4, condition *i)* guarantees the C^k-solvability in U_j, and by Theorem 3.13 condition *ii)* guarantees the global C^k-solvability.

To prove the necessity of condition *i)* we note that as we have already seen the restriction of (3.2.4) to U_j is equivalent to the equation (3.2.7) with $\lambda = \lambda_j$ and $\mu = \mu_j$. Therefore

$$(\lambda_j^s - \mu_j)\psi^{(s)}(0) = \tilde{\gamma}^{(s)}(0), \quad s = 0, \dots, k,$$

and since $\tilde{\gamma}^{(s)}(0)$ may be arbitrary numbers, we conclude that $\lambda_j^s \neq \mu_j$, $s = 0, \dots, k$; $j = 1, \dots, n$. To complete the proof it remains to check the necessity of inequalities $|\mu_j| \neq \lambda_j^k$ for all j.

Since $\mu_j \neq \lambda_j^k$, the assumption $|\mu_j| = \lambda_j^k$ for some j, $1 \leq j \leq n$, implies $\mu_j < 0$. Let us construct a C^k-function γ such that (3.2.4) has no C^k-solutions.

To be definite, we assume $\lambda_j < 1$ and consider the equation

$$\psi(\lambda_j t) - \mu_j \psi(t) = h(t), \quad t \in \mathbb{R},$$

with

$$h(t) = \int_0^t \frac{(t - s)^{k-1}}{(k-1)!} \frac{\cos(\pi(\ln \lambda_j)^{-1} \ln |s|)}{1 + |\ln |s||} \, ds.$$

If ψ is its local C^k-solution at $t = 0$, then the derivative $\psi^{(k)}$ satisfies the equation

$$\lambda_j^k \psi^{(k)}(\lambda_j t) - \mu_j \psi^{(k)}(t) = h^{(k)}(t), \quad t \in \mathbb{R}.$$

Iterating it we find

$$\psi^{(k)}(t) = (-1)^n \psi^{(k)}(\lambda_j^n t) - \frac{1}{\mu_j} \sum_{p=0}^{n-1} (-1)^p \, h^{(k)}(\lambda_j^p t),$$

and the sum here must be bounded as $n \to \infty$. On the other hand,

$$(-1)^p h^{(k)}(\lambda_j^p t) = \frac{\cos(\pi(\ln \lambda_j)^{-1} \ln |t|)}{1 + |p \ln \lambda_j + \ln |t||}, \quad p = 1, 2, \dots,$$

contradicting the boundedness of the sum.

Therefore if a C^k-function $\gamma(x)$ coincides with the function

$$\gamma_0(x) = S_j^{-1}(x) h(\Psi_j(x))$$

in a neighborhood of x_j, then (3.2.4) has no local C^k-solution at x_j. □

Example 3.10. The function

$$F(x) = x^3 + \rho x, \quad \rho \in (0,1),\ x \in \mathbb{R},$$

defines an analytic diffeomorphism of the real line,

$$\operatorname{Fix} F = \left\{-\sqrt{1-\rho},\ 0,\ \sqrt{1-\rho}\right\}.$$

Here

$$\lambda_1 = \lambda_3 = 3 - 2\rho, \quad \lambda_2 = \rho,$$

and therefore all fixed points are hyperbolic. Given a C^k-function $a(x)$, $a(x) \neq 0$, $x \in \mathbb{R}$, set

$$\mu_1 = a(-\sqrt{1-\rho}), \quad \mu_2 = a(0), \quad \mu_3 = a(\sqrt{1-\rho}).$$

It follows from Theorem 3.14 that the equation

$$\varphi(x^3 + \rho x) - a(x)\varphi(x) = \gamma(x)$$

has a C^k-solution for every C^k-function γ if and only if

$$\mu_j \neq \lambda_j^s,\ s = 0, 1, \ldots, k; \quad |\mu_j| \neq \lambda_j^k,\ j = 1, 2, 3,$$

and

$$k < \max\left(\frac{\ln|\mu_1|}{\ln(3-2\rho)}, \frac{\ln|\mu_2|}{\ln\rho}\right), \quad k < \max\left(\frac{\ln|\mu_2|}{\ln\rho}, \frac{\ln|\mu_3|}{\ln(3-2\rho)}\right).$$

In particular, the equation with a constant coefficient $a(x) = \mu$ has a C^k-solution for every C^k-function γ if and only if $\rho^k > |\mu|$ in the case $|\mu| < 1$ and $(3-2\rho)^k < |\mu|$ otherwise.

3.4 Comments and references

3.1. Theorem 3.1 for analytic functions is a one-dimensional version of a multidimensional result first stated in [29]. Complete proof may be found in [30]. The condition $F'(x_0) > 0$, which means that F preserves the orientation, is redundant in dimensions greater than 1.

The Lipschitz condition in Theorem 3.2 is redundant, see [53]. In the present form Theorem 3.3 on the local C^∞-solvability is exposed here for the first time.

3.2. To the best of our knowledge an analogue of Theorem 3.5 for general analytic mappings $G(x,y) = (F(x), g(x,y))$ with $\operatorname{Fix} F = \emptyset$ is not known. The proof of existence of a global analytic solution (Theorem 3.6) is published here for the first time.

3.3. The cohomological approach to a study of linear equations first appeared in [21], see also [32].

A continuous version of a generalized Abel equation has the form

$$(v\varphi)(x) = h(x, \varphi(x)), \qquad x \in M, \tag{3.4.1}$$

where $v = v(x)\partial/\partial x$ is a vector field on M, and $h(x, y)$ is a function on $M \times \mathbb{R}$. It may be treated using the methods of Chapter 3. In particular, the following analogue of Theorem 3.3 holds:

Let v be a C^∞-vector field, non-flat at its singular point x_0, and let $h \in C^\infty$. If equation (3.4.1) *has a formal solution $\hat{\varphi}_0$ at x_0, then there exists a local C^∞-solution φ whose Taylor series at x_0 is equal to $\hat{\varphi}_0$.*

Chapter 4

Equations with Several Transformations of Argument

In this section we consider equations of the type

$$g(x, \varphi(F_1(x)), \ldots, \varphi(F_n(x))) = 0, \quad x \in \mathbb{R}. \tag{4.0.1}$$

Properties of this equation depend on a joint dynamical behavior of the mappings $F_1, \ldots, F_n$, which may be rather complicated. Here we consider some properties which influence the solvability of (4.0.1). For the sake of simplicity we assume that the function $g(x, y_1, \ldots, y_n)$ in equation (4.0.1) is defined for all $(x, y_1, \ldots, y_n) \in \mathbb{R}^{n+1}$.

We start with a problem of local solvability at a closed subset $V \subset \mathbb{R}$.

4.1 Local solvability

In Section 3.1 we defined a notion of a local solution to a generalized Abel equation. Now we extend this notion to solutions of equation (4.0.1).

Definition 4.1. Equation (4.0.1) is said to be *locally C^k-solvable in a neighborhood of a closed subset $V \subset \mathbb{R}$* (or just *at V*) if there exists a C^k-function φ defined in a connected neighborhood of the set $\bigcup_{j=1}^{n} F_j(V)$ and a neighborhood U of V such that

$$g(x, \varphi(F_1(x)), \ldots, \varphi(F_n(x))) = 0, \quad x \in U.$$

The function φ is called a *local C^k-solution at V*.

We will prove an analogue of Theorem 3.2 for an equation

$$\varphi(x) = h(x, \varphi(H_1(x)), \dots, \varphi(H_n(x))) \qquad (4.1.1)$$

where $h(x, y_1, \dots, y_n)$ and $H_j(x)$, $j = 1, 2, \dots, n$, are C^k-functions, which is a particular case of (4.0.1) with $F_1(x) \equiv x$. We restrict ourselves to the case of either $k < \infty$ or $k = \infty$ only, excluding $k = \aleph$. If $k < \infty$, we assume, in addition, that all partial derivatives of order k satisfy the Lipschitz condition: for every $N > 0$ there exists a constant $c = c(N)$ such that

$$\left| \frac{\partial^k h(x, y_1', \dots, y_n')}{\partial y_1^{p_1} \dots \partial y_n^{p_n}} - \frac{\partial^k h(x, y_1'', \dots, y_n'')}{\partial y_1^{p_1} \dots \partial y_n^{p_n}} \right| \le c \sum_{j=1}^{n} |y_j' - y_j''| \qquad (4.1.2)$$

for all $p_j \ge 0$, $\sum p_j = k$, all $y_1', \dots, y_n', y_1'', \dots, y_n''$ and x such that

$$\sum_{j=1}^{n} |y_j'| \le N, \quad \sum_{j=1}^{n} |y_j''| \le N, \quad x \in U.$$

In the case $k = 0$ we assume $c(N) \le \rho < 1$ for all N.

Theorem 4.1. *Let V be a finite closed interval and let U be its neighborhood such that*

$$\sum_{j=1}^{n} \left| \frac{\partial h(x, y_1, \dots, y_n)}{\partial y_j} \right| |H_j'(x)|^p \le \rho < 1 \qquad (4.1.3)$$

for all $x \in U$, all $(y_1, \dots, y_n) \in \mathbb{R}^n$ and $p = 0, \dots, k$. Then (4.1.1) is locally C^k-solvable at V.

Proof. We have to find a function $\varphi(x)$ defined in a connected neighborhood of the set

$$\bigcup_{j=1}^{n} H_j(V) \bigcup V$$

and satisfying (4.1.1) in a neighborhood of V. Denote by $\tau : \mathbb{R} \to [0, 1]$ a C^k-function equal to 1 in a neighborhood of V and vanishing outside U. Then the operator

$$T\varphi(x) = \tau(x) h(x, \varphi(H_1(x)), \dots, \varphi(H_n(x)))$$

is well defined in the space $\mathcal{E}_0(U)$ of C^k-functions φ vanishing outside U. It is sufficient to prove the C^k-solvability of equation

$$\varphi = T\varphi, \quad \varphi \in \mathcal{E}_0(U).$$

To this end we denote by $B(M_0, \dots, M_k)$ the set of functions $\varphi \in \mathcal{E}_0(U)$ such that

$$|\varphi^{(p)}(x)| \le M_p, \quad p = 0, 1, \dots, k,$$

and easily check that $\varphi \in B(M_0, \ldots, M_k)$ implies

$$|(T\varphi)^{(p)}(x)| \leq \rho M_p + R_p(M_0, \ldots, M_{p-1}), \quad p = 0, 1, \ldots, k,$$

where $\rho < 1$ and R_p is a polynomial determined by the functions h, $H_1 \ldots H_n$, which does not depend on φ.

In particular, $\varphi \in B(M_0, \ldots, M_k)$ implies

$$|T\varphi(x)| \leq \rho M_0 + R_0, \quad x \in \mathbb{R},$$

where

$$R_0 = \max_{x \in U} |h(x, 0, \ldots, 0)|.$$

We choose $M_0, M_1, \ldots, M_k$ such that

$$M_0 > \frac{R_0}{1 - \rho}; \quad M_p > \frac{R_p(M_0, \ldots, M_{p-1})}{1 - \rho}, \quad p = 1, 2, \ldots, k,$$

and find that the set $B(M_0, \ldots, M_k)$ is invariant with respect to T. If $k = \infty$, then this set is convex and compact in $\mathcal{E}_0(U)$ and the proof is completed by using the Leret–Schauder theorem [42].

For $k < \infty$ and $\varphi_1, \varphi_2 \in B(M_0, \ldots, M_k)$ we set

$$d_p(\varphi_1, \varphi_2) = \max_{x \in \mathbb{R}} |\varphi_1^{(p)}(x) - \varphi_2^{(p)}(x)|, \qquad p = 0, \ldots, k.$$

Then estimates (4.1.3) yield

$$|T\varphi_1(x) - T\varphi_2(x)| \leq \rho d_0(\varphi_1, \varphi_2)$$

and

$$|(T\varphi_1)^{(p)}(x) - (T\varphi_2)^{(p)}(x)| \leq \rho d_p(\varphi_1, \varphi_2) + c_1 \sum_{s=0}^{p-1} d_s(\varphi_1, \varphi_2),$$

with the same constant c_1 for $p = 1, \ldots, k$. We choose numbers $\alpha_0, \ldots, \alpha_{k-1}, \alpha_k = 1$ such that

$$\tilde{\rho} = \rho + \max_{1 \leq p \leq k-1} \frac{c_1}{\alpha_p} \sum_{s=p+1}^{k} \alpha_s < 1,$$

and set

$$\operatorname{dist}(\varphi_1, \varphi_2) = \sum_{p=0}^{k} \alpha_p d_p(\varphi_1, \varphi_2). \tag{4.1.4}$$

Then for $\varphi_1, \varphi_2 \in B(M_0, \ldots, M_k)$ we obtain

$$\operatorname{dist}(T\varphi_1, T\varphi_2) \leq \tilde{\rho} \operatorname{dist}(\varphi_1, \varphi_2).$$

Hence operator T is contractive in the set $B(M_0, \ldots, M_k)$ endowed with the metric (4.1.4) and there exists a function $\varphi \in B(M_0, \ldots, M_k)$ such that $\varphi = T\varphi$. □

Let us note that, as Example 5.3 below shows, the statement of Theorem 4.1 fails if V is not a bounded interval.

4.2 Extension of solutions

Here a generalization of Theorem 3.7 will be given: we describe some subsets V such that the local solvability of (4.0.1) at V implies global solvability.

4.2.1 Absorbers

We will call a closed interval $V \subset \mathbb{R}$, either finite or infinite, an *absorber* for a continuous mapping $H : \mathbb{R} \to \mathbb{R}$ if V is H-invariant and for every point $x \in \mathbb{R}$ there exists a neighborhood U and a number N such that

$$H^s(U) \subset \mathrm{Int} V, \quad s \geq N,$$

where $\mathrm{Int}\, V$ is the interior of V. It is easy to see that every absorber contains all non-wandering points of H.

Example 4.1. Let $H(x) = \lambda x$. If $\lambda \in (0,1)$, then every interval $[a,b]$, $a < 0 < b$ is an absorber. If $\lambda \in (-1,0)$, then $[a,b]$ is an absorber if and only if $a < 0 < b$ and $\lambda^{-1}b \leq a \leq \lambda b$. In the case $\lambda \notin (-1,1)$ H has no absorbers, except $V = \mathbb{R}$.

Example 4.2. Let $\mathrm{Fix}\, H = \emptyset$. If $H(x) > x$, then every interval $[a,+\infty)$ is an absorber. In the case $H(x) < x$ every interval $(-\infty, b]$ is an absorber.

If $\mathrm{Fix}\, H \neq \emptyset$, then every absorber contains the set $\mathrm{Fix}\, H$.

Further, let $\{H_j\}_1^n$ be a family of continuous mappings, $H_j : \mathbb{R} \to \mathbb{R}$, $j = 1, \ldots, n$.

Definition 4.2. A closed interval $V \subset \mathbb{R}$ is called an *absorber* for the family $\{H_j\}_1^n$ if it is H_j-invariant for $j = 1, \ldots, n$, and for every point $x \in \mathbb{R}$ there exists its neighborhood U and a number N such that

$$H_{j_1} H_{j_2} \ldots H_{j_s}(U) \subset \mathrm{Int}\, V, \quad s \geq N.$$

It is evident that if V is an absorber for a family $\{H_j\}_1^n$, then it is an absorber for every mapping H_j.

Example 4.3. Let $H_j(x) = \alpha_j x + \beta_j$, $j = 1, \ldots, n$, be a family of affine transformations. If $\alpha_j \notin [-1,1]$ at least for one index j, then the family has only a trivial absorber $V = \mathbb{R}$. The same is true if $\alpha_j = -1$ at least for some j. If $|\alpha_j| < 1$ for some j, then

$$x_j = \frac{\beta_j}{1 - \alpha_j}$$

is the unique fixed point for H_j and every absorber for the family $\{H_j\}_1^n$ must contain all these points. In the case $\alpha_j \in (0,1)$ for every $j = 1, \ldots n$, the latter condition is sufficient: every interval $[a,b]$ such that

$$a < \min_j x_j, \quad b > \max_j x_j$$

is an absorber. If $\alpha_s \in (-1,0)$ for some s and for every such s the additional condition

$$H_s^{-1}(b) \leq a \leq H_s(b)$$

is fulfilled, then the interval $[a,b]$ is an absorber for $\{H_j\}_1^n$ as well.

In addition, if $\alpha_j = 1$ for some j, $1 \leq j \leq n$, then the family has a non-trivial absorber if and only if all α_j are positive and all β_j corresponding to $\alpha_j = 1$ are not equal to zero and have the same sign. If this holds, then, depending on that sign either every interval $[a,+\infty]$, $a < \min_j x_j$ or every interval $(-\infty,b]$, $b > \max_j x_j$, is an absorber for $\{H_j\}_1^n$.

4.2.2 Extension of solutions from an absorber

Given C^k-functions $g(x,y_1,\ldots,y_n)$, $F_1(x),\ldots,F_n(x)$ and an index q, $1 \leq q \leq n$, we say that equation (4.0.1) is *invertible with respect to* y_q, if the mapping

$$(x,y_1,\ldots,y_n) \rightarrow (F_q(x),y_1,\ldots,y_{q-1},g(x,y_1,\ldots,y_n),y_{q+1},\ldots,y_n)$$

is a C^k-diffeomorphism of $\mathbb{R}^{n+1}$.

In other words this means that $F_q : \mathbb{R} \rightarrow \mathbb{R}$ is a C^k-diffeomorphism and there exists a C^k-function $h(x,y_1,\ldots,y_n)$ such that

$$g(F_q^{-1}(x),\, y_1,\ldots,y_{q-1},h(x,y_1,\ldots,y_n),y_{q+1},\ldots,y_n) = y_q$$

and

$$h(F_q(x),\, y_1,\ldots,y_{q-1},g(x,y_1,\ldots,y_n),y_{q+1},\ldots,y_n) = y_q.$$

In particular, if

$$g(x,y_1,\ldots,y_n) = a(x)y_q + g_1(x,y_1,\ldots,y_{q-1},y_{q+1},\ldots,y_n)$$

is a linear function in y_q, then equation (4.0.1) is invertible with respect to y_q if and only if $F_q : \mathbb{R} \rightarrow \mathbb{R}$ is a C^k-diffeomorphism and $a(x) \neq 0$, $x \in \mathbb{R}$.

Theorem 4.2. *Let $g(x,y_1,\ldots,y_n)$, $F_1(x),\ldots,F_n(x)$ be C^k-functions, let equation* (4.0.1) *be invertible with respect to y_q, and let V be an absorber for the family $\{F_q^{-1}F_j\}_{j\neq q}$. Then for every local C^k-solution φ_0 at V there exists a global C^k-solution of* (4.0.1) *coinciding with φ_0 on the subset $F_q(V)$.*

Proof. We assume $q = 1$ and set $\varphi(F_1(x)) = \psi(x)$. Then instead of (4.0.1) we obtain the equation

$$g(x,\psi(x),\psi(F_1^{-1}F_2(x)),\ldots,\psi(F_1^{-1}F_n(x))) = 0, \tag{4.2.1}$$

and the function $\psi_0(x) = \varphi_0(F_1(x))$ is its local C^k-solution in a neighborhood of V. Let the C^k-mapping

$$(x, y_1, \ldots, y_n) \to (F_1^{-1}(x), h(x, y_1, \ldots, y_n), y_2 \ldots y_n)$$

be inverse to the mapping

$$(x, y_1, \ldots, y_n) \to (F_1(x), g(x, y_1, \ldots, y_n), y_2, \ldots, y_n).$$

Consider the equation

$$\psi(x) = h(F_1(x), 0, \psi(F_1^{-1}F_2(x)), \ldots, \psi(F_1^{-1}F_n(x))). \tag{4.2.2}$$

Every C^k-solution of (4.2.2) is a solution of (4.2.1), and conversely, every solution of (4.2.1) is a solution of (4.2.2).

Let now $H_j = F_1^{-1}F_j$, $j = 2, \ldots, n$, and let $H_j^{-1}(W)$ be the pre-image of a set W with respect to H_j. To construct a global C^k-solution of (4.0.1) coinciding with φ_0 on the set $F_1(V)$ we define

$$V_0 = V, \quad V_1 = \bigcap_{j=2}^{n} H_j^{-1}(V_0),$$

and

$$\psi_1(x) = h(F_1(x), 0, \psi_0(H_2(x)), \ldots, \psi_0(H_n(x))).$$

Since $\psi_0(x)$ is defined in a neighborhood of V_0, the function $\psi_1(x)$ is defined in a neighborhood of V_1, and since $\psi_0(x)$ satisfies (4.2.2) at V_0, the function $\psi_1(x)$ coincides with $\psi_0(x)$ on V_0 and satisfies (4.2.2) at V_1. It is evident that V_1 is an absorber for the family $\{H_j\}_2^n$ and $V_1 \supset V_0$. Now we set recursively

$$V_{m+1} = \bigcap_{j=2}^{n} H_j^{-1}(V_m), \quad m = 0, 1, \ldots$$

and

$$\psi_{m+1}(x) = h(F_1(x), 0, \psi_m(H_2(x)), \ldots, \psi_m(H_n(x))),$$

and obtain an increasing sequence $V_0 \subset V_1 \subset \cdots \subset V_m \ldots$ of absorbers and the sequence $\psi_0(x), \psi_1(x), \ldots, \psi_m(x), \ldots$ of functions such that $\psi_m(x)$ satisfies (4.2.2) at V_m and coincides with $\psi_{m-1}(x)$ in some neighborhood of V_{m-1}. Since $V = V_0$ is an absorber, we have

$$\bigcup_{m=0}^{\infty} V_m = \mathbb{R},$$

and therefore $\psi(x) \equiv \lim_{m\to\infty} \psi_m(x)$ is a C^k-solution of (4.2.2) on $\mathbb{R}$, $\psi(x) = \psi_0(x)$ for $x \in V = V_0$. The function $\varphi(x) = \psi(F_1^{-1}(x))$ is a C^k-solution of (4.0.1) and $\varphi(x) = \varphi_0(x)$, $x \in F_1(V)$. □

4.2.3 Extension from intersection of absorbers. Decomposition method

In this section we study a possibility of extension of a local solution from the intersection of several absorbers.

Lemma 4.1. *Given closed subsets* $V_1, \ldots, V_n$ *and a* C^k*-function* $\gamma : \mathbb{R} \to \mathbb{R}$ *vanishing in a neighborhood of the set* $V = \bigcap_j V_j$, *there exist* C^k*-functions* $\gamma_j : \mathbb{R} \to \mathbb{R}$ *vanishing in neighborhoods of respective sets* V_j *such that*

$$\gamma = \sum_{j=1}^{n} \gamma_j.$$

Proof. Let $\gamma|_U = 0$ where U is a neighborhood of V. Then the subsets $V_j \setminus U$ are closed and their intersection is empty. Set

$$U_j = \mathbb{R} \setminus (V_j \setminus U), \quad j = 1, \ldots, n.$$

Then $\{U_j\}$ is an open covering of $\mathbb{R}$ and there exist C^k-functions τ_j such that

$$\tau_j(x) = 0, \quad x \notin \overline{U}_j, \qquad \sum_{j=1}^{n} \tau_j(x) = 1, \ x \in \mathbb{R}.$$

Setting $\gamma_j(x) = \tau_j(x)\gamma(x)$ we obtain a required decomposition. □

Theorem 4.3. *Let* $g(x, y_1, \ldots, y_n)$ *and* $F_1(x), \ldots, F_n(x)$ *be* C^k*-functions, let equation* (4.0.1) *be invertible with respect to* $y_{q_1}, \ldots, y_{q_s}$ *and let* $V_1, \ldots, V_s$ *be absorbers for the families*

$$\{ F_{q_1}^{-1} F_j \}_{j \neq q_1}, \ \ldots, \ \{ F_{q_s}^{-1} F_j \}_{j \neq q_s},$$

respectively. Then for every local C^k*-solution* φ_0 *of* (4.0.1) *at* $V = V_1 \bigcap \cdots \bigcap V_s$ *there exists a global* C^k*-solution coinciding with* φ_0 *on the set* $F_{q_1}(V_1) \bigcap \cdots \bigcap F_{q_s}(V_s)$.

Proof. Without loss of generality we assume that φ_0 is a C^k-function defined on the whole real line and satisfying equation (4.0.1) in a neighborhood of V. Let us write the equation in the form

$$g_0(x, \psi(F_1(x)), \ldots, \psi(F_n(x))) = \gamma(x) \tag{4.2.3}$$

where

$$\begin{aligned} \gamma(x) &= -g(x, \varphi_0(F_1(x)), \ldots, \varphi_0(F_n(x))), \\ g_0(x, y_1, \ldots, y_n) &= g(x, y_1 + \varphi_0(F_1(x)), \ldots, y_n + \varphi_0(F_n(x))) \\ &\quad - g(x, \varphi_0(F_1(x)), \ldots, \varphi_0(F_n(x))) \end{aligned}$$

and ψ is a new unknown function. If x lies in a neighborhood of V, then $g_0(x, 0, \dots, 0) = 0$ and $\gamma(x) = 0$. Hence the function $\psi_0 = 0$ is a local solution of (4.2.3) at V. Lemma 4.1 yields a decomposition $\gamma(x) = \sum_{j=1}^{s} \gamma_j(x)$ with some C^k-functions γ_j vanishing in respective neighborhoods of V_j. According to Theorem 4.2 the equation

$$g_0(x, \psi(F_1(x)), \dots, \psi(F_n(x))) = \gamma_1(x)$$

has a global C^k-solution ψ_1 vanishing on V_1. Now we set

$$g_1(x, y_1 \dots, y_n) = g_0(x, y_1 + \psi_1(F_1(x)), \dots, y_n + \psi_1(F_n(x))) - \gamma_1(x)$$

and conclude that the function $\psi = 0$ is a local solution at V_2 of the equation

$$g_1(x, \psi(F_1(x)), \dots, \psi(F_n(x))) = \gamma_2(x).$$

The same Theorem 4.2 states that there exists a global C^k-solution ψ_2 of this equation vanishing on V_2. Using induction we set

$$\begin{aligned} &g_m(x, y_1, \dots, y_n) \\ &= g_{m-1}(x, y_1 + \psi_{m-1}(F_1(x)), \dots, y_n + \psi_{m-1}(F_n(x))) - \gamma_{m-1}(x) \end{aligned}$$

where $m \leq s$ and ψ_{m-1} is a C^k-solution of the equation

$$g_{m-1}(x, \psi(F_1(x)), \dots, \psi(F_n(x))) = \gamma_{m-1}(x),$$

vanishing on V_{m-1}. Since both functions $g_m(x, 0, \dots, 0)$ and $\gamma_m(x)$ vanish on V_m, there exists a global C^k-solution ψ_m of the equation

$$g_m(x, \psi(F_1(x)), \dots, \psi(F_n(x))) = \gamma_m(x)$$

vanishing on V_m. The sum

$$\varphi(x) = \psi_1(x) + \dots + \psi_s(x) + \varphi_0(x)$$

is a global solution of (4.0.1) coinciding with the function φ_0 on the set $F_{q_1}(V_1) \bigcap \dots \bigcap F_{q_s}(V_s)$. □

Corollary 4.1. *Let all conditions of Theorem 4.3 be fulfilled and let the set $\bigcap_{j=1}^{s} V_j$ be empty. Then equation (4.0.1) has a global C^k-solution.*

If, in addition, $W = \bigcap_{j=1}^{s} F_{q_j}(V_j) \neq \emptyset$, then for every C^k-function φ_0 there exists a global C^k-solution of (4.0.1) coinciding with φ_0 in a neighborhood of W.

Combining Theorems 4.1 and 4.3 we obtain conditions of the global solvability in the case of non-empty set $V = \bigcap_{j=1}^{s} V_j$.

Corollary 4.2. *Let $g(x, y_1, \dots, y_n)$ and $F_1(x), \dots, F_n(x)$ be C^k-functions and the following conditions hold:*

i) Equation (4.0.1) *is invertible with respect to variables $y_q, y_{q_1}, \dots, y_{q_s}$ and the function $h(x, y_1, \dots, y_n)$ satisfies*

$$g(F_q^{-1}(x), y_1, \dots, y_{q-1}, h(x, y_1, \dots, y_n), y_{q+1}, \dots, y_n) \equiv y_q;$$

ii) Intervals $V_1, \dots, V_s$ are absorbers for the families

$$\{ F_{q_1}^{-1} F_j \}_{j \neq q_1}, \dots, \{ F_{q_s}^{-1} F_j \}_{j \neq q_s},$$

respectively, and the set $V = V_1 \cap \dots \cap V_s$ is a closed bounded interval;

iii) The C^k-function

$$\tilde{h}(x, y_1, \dots, y_{q-1}, y_{q+1}, \dots, y_n) \equiv h(F_q(x), y_1, \dots, y_{q-1}, 0, y_{q+1}, \dots, y_n)$$

satisfies the Lipschitz condition (4.1.2) *and estimates* (4.1.3) *are valid with $H_j = F_q^{-1} F_j$.*

Then equation (4.0.1) *has a global C^k-solution.*

Proof. Indeed, because of the invertibility with respect to y_q, equation (4.0.1) is equivalent to

$$\psi(x) = \tilde{h}(x, \psi(H_1(x)), \dots, \psi(H_{n-1}(x))).$$

Now it is sufficient to note that by Theorem 4.1 conditions (4.1.2) and (4.1.3) guarantee a local solvability at V and Theorem 4.3 implies the global solvability. □

As we have already noted, the invertibility with respect to y_q for a linear equation

$$\sum_{j=1}^{n} a_j(x) \varphi(F_j(x)) = \gamma(x) \tag{4.2.4}$$

means that F_q is a diffeomorphism and $a_q(x) \neq 0$, $x \in \mathbb{R}$. The function $\tilde{h}$ has the form

$$\tilde{h}(x, y_1, \dots, y_n) = -\sum_{j \neq q} \frac{a_j(x)}{a_q(x)} y_j + \frac{\gamma(x)}{a_q(x)}.$$

The Lipschitz condition is trivially fulfilled, estimates (4.1.2) are simplified and we obtain the following statement.

Corollary 4.3. *Let* $a_j(x) \in C^k$, $F_j(x) \in C^k$ *and the following conditions hold:*

i) The mappings $F_q, F_{q_1}, \dots, F_{q_s}$ *are* C^k*-diffeomorphisms for some indices* $q, q_1, \dots, q_s$;

ii) The functions $a_q(x)$, $a_{q_1}(x), \dots, a_{q_s}(x)$ *do not vanish on the real line;*

iii) Intervals $V_1, \dots, V_s$ *are absorbers for the families* $\{F_{q_1}^{-1}F_j\}_{j\neq q_1}, \dots$, $\{F_{q_s}^{-1}F_j\}_{j\neq q_s}$, *respectively, and the set* $V = \bigcap_{j=1}^{s} V_j$ *is a closed bounded interval;*

iv) For all $x \in V$ *an estimate*

$$\sum_{j\neq q}\left|\frac{a_j(x)}{a_q(x)}\right| \, |(F_q^{-1}F_j(x))'|^p \le \rho < 1, \quad p = 0, 1, \dots, k, \tag{4.2.5}$$

holds.

Then equation (4.2.4) *has a global* C^k*-solution* $\varphi(x)$ *on the real line.*

4.3 Examples

Let us consider equation (4.0.1) with some special families of mappings $\{F_j\}$.

1. Let $n = 2$. If equation (4.0.1) is invertible with respect to y_1, then it is reduced to the generalized Abel equation

$$\psi(x) = h(x, \psi(F_1^{-1}F_2(x))).$$

Assume that Fix $(F_1^{-1}F_2) = \emptyset$. Then either $F_1^{-1}F_2(x) < x$, and every interval $(-\infty, b]$ is an absorber or $F_1^{-1}F_2(x) > x$ and every interval $[a, +\infty)$ is an absorber for $F_1^{-1}F_2$. Every local solution of (4.0.1) at such an interval is extended to a global solution.

Let now Fix $(F_1^{-1}F_2) \neq \emptyset$. Set

$$x_- = \min \operatorname{Fix}(F_1^{-1}F_2), \quad x_+ = \max \operatorname{Fix}(F_1^{-1}F_2).$$

If we assume that $x_- > -\infty$, the interval $[x_-, +\infty)$ is invariant with respect to $F_1^{-1}F_2$ and x_- attracts from the left, then every interval $[a, +\infty)$, $a < x_-$, is an absorber. Similarly, if $x_+ < +\infty$, the interval $(-\infty, x_+]$ is invariant with respect to $F_1^{-1}F_2$ and x_+ attracts from the right, then every interval $(-\infty, b]$, $b > x_+$, is an absorber.

If the interval $[x_-, x_+]$ is invariant with respect to $F_1^{-1}F_2$, x_- attracts from the left and x_+ attracts from the right, then every interval

$$[a, b], \quad a < x_- \le x_+ < b, \tag{4.3.1}$$

is an absorber. A local solvability at such an interval implies the global one.

If x_- repels from the left and x_+ repels from the right, then the mapping $F_1^{-1}F_2$ has no non-trivial absorbers. But if the equation is invertible also with respect to y_2, then every interval (4.3.1) is an absorber either for $F_1^{-1}F_2$ or for $F_2^{-1}F_1$, or is an intersection of an absorber for the former with an absorber for the latter. According to Theorem 4.3 we conclude that:

If $n = 2$ and equation (4.0.1) is invertible with respect to y_1 and y_2, then everyone of its local C^k-solutions in a neighborhood of $[x_-, x_+]$ may be continued as a global C^k-solution.

2. Now we consider (4.0.1) with $n = 3$ and assume that the equation is invertible with respect to variables y_1, y_2, y_3. Then without loss of generality we can consider the equation

$$g(x, \psi(x), \psi(F_1(x)), \psi(F_2(x))) = 0. \tag{4.3.2}$$

Suppose that F_1 and F_2 preserve the orientation and $F_1(x) < F_2(x)$, $x \in \mathbb{R}$. Set

$$x_- = \min \operatorname{Fix} F_1; \quad x_+ = \max \operatorname{Fix} F_2.$$

We accept $x_- = +\infty$ if $\operatorname{Fix} F_1 = \emptyset$ and $x_+ = -\infty$ in the case $\operatorname{Fix} F_2 = \emptyset$.

Let the points x_- and x_+ be repellers for F_1 and F_2, respectively. Then every interval $[a, +\infty)$, $a < x_-$, is an absorber for the family $\{F_1^{-1}, F_1^{-1}F_2\}$ and every interval $(-\infty, b]$, $b > x_+$, is an absorber for the family $\{F_2^{-1}, F_2^{-1}F_1\}$. It follows from Theorem 4.3 that:

If F_1 and F_2 preserve the orientation, $F_1(x) < F_2(x)$ and equation (4.3.2) is locally C^k-solvable at the interval $[x_-, x_+]$, then it has a global C^k-solution. In particular, according to Corollary 4.1, if $x_+ < x_-$, then equation (4.3.2) has a global C^k-solution.

Example 4.4. Consider equation (4.3.2) with

$$F_1(x) = x^3 + \alpha x, \quad F_2(x) = x^3 + \alpha x + \beta, \ \alpha > 1, \ \beta > 0. \tag{4.3.3}$$

The mappings F_1 and F_2 are analytic diffeomorphisms,

$$\operatorname{Fix} F_1 = \{0\}, \quad \operatorname{Fix} F_2 = \{z\}$$

where $z < 0$ is the unique real root of the equation

$$x^3 + (\alpha - 1)x + \beta = 0.$$

Since

$$F_1'(x) = F_2'(x) = 3x^2 + \alpha > 1,$$

the points $x_- = 0$ and $x_+ = z$ are repellers for F_1 and F_2, respectively. Since $z < 0$, the interval $[x_-, x_+]$ is empty. Hence if equation (4.3.2) with mappings (4.3.3) is invertible with respect to y_2 and y_3, then it is globally C^k-solvable.

For instance, given C^k-functions $a_1(x) \neq 0$, $a_2(x) \neq 0$ and $h(x,y)$, the equation

$$a_1(x)\varphi(x^3 + \alpha x) + a_2(x)\varphi(x^3 + \alpha x + \beta) = h(x, \varphi(x)),$$

has a global C^k-solution.

3. A linear mapping

$$F(x) = \lambda x + \mu$$

with $\lambda \neq 1$ has the unique fixed point

$$x_0 = \frac{\mu}{1 - \lambda}$$

which either attracts or repels depending on whether $|\lambda| < 1$ or $|\lambda| > 1$, respectively. If $0 < \lambda < 1$, then the sequence $\{F^n(x)\}_{n=0}^{\infty}$ is monotonic and converges to x_0 as $n \to +\infty$. If $-1 < \lambda < 0$, then $\{F^{2n}(x)\}_{n=0}^{\infty}$ and $\{F^{2n+1}(x)\}_{n=0}^{\infty}$ are monotonic sequences converging to x_0 from opposite directions as $n \to +\infty$. In any case, if $|\lambda| < 1$, then every neighborhood of x_0 contains an absorber.

Let us now consider equation (4.0.1) with affine transformations

$$F_j(x) = \lambda_j x + \mu_j, \quad \lambda_j \neq 0, \; j = 1, \dots, n, \tag{4.3.4}$$

such that

$$|\lambda_1| = \cdots = |\lambda_s| > |\lambda_j|, \quad j \geq s + 1, \tag{4.3.5}$$

for some s, $1 \leq s \leq n$. We consider two cases: either $s = 1$, or all λ_j's have the same sign. Otherwise all absorbers for the families $\{F_q^{-1}F_j\}_{j \neq q}$, $q = 1, \dots, n$, coincide with $\mathbb{R}$.

Let us start with the case $s = 1$ and show that the family $\{F_1^{-1}F_j\}_{j \neq 1}$ has compact absorbers. Indeed, the points

$$x_j = \frac{\mu_j - \mu_1}{\lambda_1 - \lambda_j}, \quad j = 2, \dots, n,$$

are fixed for $H_j = F_1^{-1}F_j$, $j = 2, \dots, n$. We set

$$z_- = \min_{2 \leq j \leq n} x_j, \quad z_+ = \max_{2 \leq j \leq n} x_j.$$

If all λ_j have the same sign, then $0 < \lambda_j \lambda_1^{-1} < 1$ and every interval $[a, b]$ with $a < z_- \leq z_+ < b$ is an absorber for $\{H_j\}_{j=2}^{n}$. Otherwise there are λ_j with different signs and a segment $[a, b]$ is an absorber under the additional condition

$$\max_{2 \leq j \leq n;\; \lambda_j \lambda_1 < 0} H_j^{-1}(b) \leq a \leq \min_{2 \leq j \leq n;\; \lambda_j \lambda_1 < 0} H_j(b).$$

In order to find the minimal interval satisfying these conditions, let us note that the point

$$x_{ij} = \frac{\lambda_i(\mu_j - \mu_1) + \lambda_1(\mu_i - \mu_1)}{\lambda_1^2 - \lambda_i\lambda_j}$$

is the unique fixed point of the mapping H_iH_j with $\lambda_i\lambda_1 < 0$, $\lambda_j\lambda_1 < 0$. Since $(H_iH_j(x))' = \lambda_i\lambda_j\lambda_1^{-2} < 1$, this point is attractive. Therefore $H_i^{-1}(b) \le H_j(b)$ if and only if $b \ge x_{ij}$ and

$$\max_{2\le j\le n;\ \lambda_j\lambda_1<0} H_j^{-1}(b) \le \min_{2\le j\le n;\ \lambda_j\lambda_1<0} H_j(b)$$

if and only if $b \ge z_1 = \max_{i,j} x_{ij}$ for all i, j such that $\lambda_i\lambda_1 < 0$, $\lambda_j\lambda_1 < 0$. Since $a < z_-$, we must require $z_- \ge H_j^{-1}(b)$ for all j such that $\lambda_j\lambda_1 < 0$, i.e., $b \ge z_2 = \max_j H_j(z_-)$. Now we set

$$\tilde z_+ = \max(z_1, z_2, z_+), \quad \tilde z_- = \min\{\min_{2\le j\le n;\ \lambda_1\lambda_j<0}\{H_j(\tilde z_+)\}, z_-\}$$

and find that the inequalities

$$\max_{2\le j\le n;\ \lambda_1\lambda_j<0} H_j^{-1}(\tilde z_+) \le \tilde z_- \le \min_{2\le j\le n;\ \lambda_1\lambda_j<0} H_j(\tilde z_+)$$

hold. The interval $[\tilde z_-, \tilde z_+]$ is invariant with respect to the family $\{H_j\}$, every absorber must contain it and any of its neighborhoods contains an absorber. To describe all absorbers we choose an arbitrary $b > \tilde z_+$ and an arbitrary a satisfying the inequalities

$$\max_{2\le j\le n;\ \lambda_j\lambda_1<0} H_j^{-1}(b) < a < \min\{\min_{2\le j\le n;\ \lambda_j\lambda_1<0}\{H_j(b)\}, z_-\}.$$

Therefore:

If $F_j(x)$ are affine transformations of the form (4.3.4) such that (4.3.5) is valid with $s = 1$ and equation (4.0.1) is invertible with respect to y_1, then everyone of its local C^k-solutions at $[\tilde z_-, \tilde z_+]$ is extended as a global C^k-solution.

Example 4.5. Consider the equation

$$\varphi(x) = h(x, \varphi(\alpha_2 x), \varphi(\alpha_3 x + \beta_3), \dots, \varphi(\alpha_n x + \beta_n)) \tag{4.3.6}$$

with $\alpha_2 \in (-1, 0)$, $\alpha_j \in (0, 1)$, $j = 3, \dots, n$. Here

$$z_- = \min_{3\le j\le n}\left(0, \frac{\beta_j}{1 - \alpha_j}\right), \quad z_+ = \max_{3\le j\le n}\left(0, \frac{\beta_j}{1 - \alpha_j}\right),$$

$$z_1 = x_{11} = 0, \quad z_2 = \alpha_2 z_-,$$

and

$$\tilde z_+ = \max\{\alpha_2 z_-, z_+\}, \quad \tilde z_- = \min\{\alpha_2 \tilde z_+, z_-\}.$$

Corollary 4.3 implies that if $h(x, y_2, \ldots, y_n) \in C^k$, the Lipschitz condition (4.1.2) holds and an estimate

$$\sum_{j=2}^{n} \left| \frac{\partial h(x, y_2, \ldots, y_n)}{\partial y_j} \right| \le \rho < 1$$

is valid for x in a neighborhood of the interval $[\tilde{z}_-, \tilde{z}_+]$ for all $y_j \in \mathbb{R}, 1 \le j \le n$, then equation (4.3.6) has a global C^k-solution.

Let us now assume that s in (4.3.5) is an arbitrary number, $1 \le s \le n$, $\lambda_j > 0,\ j = 1, \ldots, n$, and

$$\mu_1 < \cdots < \mu_s.$$

If $s = n$, then we set $z_- = +\infty$, $z_+ = -\infty$. If $s \le n - 1$, then, as above, we set

$$z_- = \min_{j>s} \frac{\mu_j - \mu_1}{\lambda - \lambda_j}, \quad z_+ = \max_{j>s} \frac{\mu_j - \mu_s}{\lambda - \lambda_j} \tag{4.3.7}$$

where $\lambda = \lambda_1 = \cdots = \lambda_s$. Every interval $[a, +\infty)$, $a < z_-$, is an absorber for $\{F_1^{-1}F_j\}_{j \ne 1}$ and every interval $(-\infty, b]$, $b > z_+$, is an absorber for $\{F_s^{-1}F_j\}_{j \ne s}$. Therefore:

If F_j are affine transformations (4.3.4) with $\lambda_j > 0$, equation (4.0.1) is invertible with respect to y_1 and y_s and is locally C^k-solvable in a neighborhood of the interval $[z_-, z_+]$, then according to Theorem 4.3 it has a global C^k-solution.

In particular, if $s = n$, then the interval $[z_-, z_+]$ is empty. It may happen that it is empty even if $s < n$. Then we can choose $a < z_-$ and $b > z_+$ to be such that $[a, +\infty) \bigcap (-\infty, b] = \emptyset$ and arrive at an "unconditional" existence theorem. For instance, if $s = n - 1$, then

$$z_+ = \frac{\mu_n - \mu_{n-1}}{\lambda - \lambda_n} < z_- = \frac{\mu_n - \mu_1}{\lambda - \lambda_n}.$$

Example 4.6. Given C^k-functions $a_1(x) \ne 0$, $a_2(x) \ne 0$ and $h(x, y)$, the equation

$$a_1(x)\varphi(\lambda x + \mu_1) + a_2(x)\varphi(\lambda x + \mu_2) = h(x, \varphi(x))$$

has a global C^k-solution if $\lambda > 1$ and $\mu_1 \ne \mu_2$.

Example 4.7. Consider the linear equation

$$\varphi(x) = \sum_{j=1}^{s} a_j(x)\varphi(x + \beta_j) + \sum_{j=s+1}^{n} a_j(x)\varphi(\lambda_j x + \beta_j) + \gamma(x)$$

with $0 < \beta_1 < \cdots < \beta_s, \lambda_j \in (0, 1)$ and C^k-coefficients $a_j(x)$. Here $q = 1$, $F_1(x) = x$, $F_s(x) = x + \beta_s$ and hence

$$z_- = \min_{j \ge s+1} \frac{\beta_j}{1 - \lambda_j}, \quad z_+ = \max_{j \ge s+1} \frac{\beta_j - \beta_s}{1 - \lambda_j}, \quad V = [z_-, z_+].$$

Since $0 < \lambda_j < 1$, condition (4.2.5) is reduced to the inequality

$$\sum_{j=1}^{n} |a_j(x)| \le \rho < 1, \quad x \in V.$$

If the latter is valid and a_s does not vanish anywhere, then the equation has a global C^k-solution on $\mathbb{R}$ for every C^k-function γ.

The following example shows that the condition $z_+ < z_-$ is not necessary for the global solvability of (4.0.1) with every C^k-function $g(x, y_1, \dots, y_n)$, invertible with respect to variables $y_1, \dots, y_n$.

Example 4.8. Let $n = 5, F_1(x) = x, F_2(x) = x + 2, F_3(x) = \lambda x, F_4(x) = \lambda x + 1, F_5(x) = \lambda x + 2$ where $\lambda \in (0, 1)$. Then the interval $[z_-, z_+]$ is reduced to the point $z = 0$ and hence is not empty. However if a C^k-function $g(x, y_1, \dots, y_5)$ is such that the equation

$$g(x, \varphi(x), \varphi(x+2), \varphi(\lambda x), \varphi(\lambda x + 1), \varphi(\lambda x + 2)) = 0$$

is invertible with respect to y_1, y_2, y_4, then it has a global C^k-solution. Indeed, it is sufficient to construct a local C^k-solution in a neighborhood of $V = \{0\}$. To this end we choose neighborhoods U_1 of V and U_2 of $F_4(V) = \{1\}$ such that $\overline{U_1} \cap \overline{U_2} = \emptyset$. Because of the invertibility with respect to y_4 the equation

$$g(x, 0, 0, 0, \psi(\lambda x + 1), 0) = 0$$

has a global C^k-solution ψ. Every C^k-function coinciding with

$$\varphi(x) = \begin{cases} 0, & x \in U_1, \\ \psi(x), & x \in U_2, \end{cases}$$

in $U_1 \cup U_2$ is a local C^k-solution at V.

As we have noted above, non-trivial absorbers exist only if either $s = 1$ or all numbers $\lambda_1, \dots, \lambda_n$ are of the same sign. Theorem 4.3 is not applicable outside this restriction and additional conditions for the solvability arise. The simplest equation of such a type

$$\varphi(-x) + \varphi(x) = \gamma(x)$$

is solvable if and only if γ is an even function.

4.4 Difference equations in Carleman classes

The proof of Theorem 4.3 on the solvability of general equation (4.0.1) is based on a decomposition of a function into a sum of functions vanishing on corresponding absorbers. Of course, such decomposition is not possible for non-zero analytic and quasianalytic functions. Nevertheless the decomposition method may be modified using functions rapidly decreasing at infinity. Here we demonstrate such a modification for solving linear difference equations

$$T\varphi(x) \equiv \sum_{i=1}^{q} a_i(x)\varphi(x+\alpha_i) = \gamma(x), \quad \alpha_1 < \alpha_2 < \cdots < \alpha_q. \tag{4.4.1}$$

As everywhere in the present chapter we consider equations with real-valued coefficients and look for real-valued solutions, though some of results are valid for complex-valued functions.

4.4.1 Decomposition in classes $C(m_n)$

Let us remind the reader of the definition of the Carleman classes of C^∞-functions.

Given a positive sequence $\{m_n\}_0^\infty$, the class $C(m_n)$ is the set of all C^∞-functions $f : \mathbb{R} \to \mathbb{R}$ such that

$$| f^{(n)}(x) | \le c r^n m_n, \quad n = 0, 1, \ldots,$$

for any bounded interval $I \subset \mathbb{R}$ with some constants $c = c(f, I)$, $r = r(f, I)$. In what follows we assume that a sequence $\{m_n\}$ satisfies the regularity condition

$$l_n^2 \le l_{n+1} l_{n-1}, \quad l_n = m_n/n!.$$

If the condition

$$\sum^{\infty} \frac{1}{\sqrt[n]{m_n}} = \infty$$

holds, the space $C(m_n)$ is the *Carleman quasianalytic class*: every function $f \in C(m_n)$ with $f^{(k)}(a) = 0$ for some $a \in \mathbb{R}$ and $k = 0, 1, 2, \ldots$ is identically equal to 0. In particular, $C(n!)$ is the space of all functions analytic on the real line.

Given a sequence $\{m_n\}_1^\infty$, let

$$h(r) = \inf_n \frac{m_n}{n!} r^n = \inf_n l_n r^n, \quad r > 0.$$

It is proved in [40] that a function $\gamma(x)$ belongs to class $C(m_n)$ if and only if the following two equivalent conditions hold:

i) For each interval $I \subset \mathbb{R}$ there exists an extension $\hat{\gamma}(z)$ of $\gamma(x)$ into a neighborhood U of I in the complex plane such that

$$|\bar{\partial}\hat{\gamma}(z)| \equiv \left| \frac{\partial \hat{\gamma}(z)}{\partial \bar{z}} \right| \le h(Q \mid \operatorname{Im} z \mid), \quad z \in U,$$

with some constant $Q = Q(I)$. The function $\hat{\gamma}$ is called a *pseudoanalytic continuation* of γ.

ii) γ admits an integral representation of the form

$$\gamma(x) = \int_{\mathbb{C}} \frac{\varphi(\zeta)}{\zeta - x} d\xi d\eta, \ \zeta = \xi + i\eta, \quad x \in \mathbb{R},$$

where $\varphi \in L^1(\mathbb{C})$ and for each interval $I \subset \mathbb{R}$ there exists a neighborhood U of I in the complex plane such that

$$| \varphi(\zeta) | \leq h(Q \mid \mathrm{Im}\zeta \mid), \quad \zeta \in U,$$

with some constant $Q = Q(I)$.

Let $\hat{a}$ and $\hat{b}$ be pseudoanalytic continuations of functions a and b, respectively, belonging to $C(m_n)$. Then the product $\hat{a}\hat{b}$ is a pseudoanalytic continuation of ab and hence $ab \in C(m_n)$. Therefore every class $C(m_n)$ forms an $\mathbb{R}$-algebra. More generally, given a real-analytic function $f(x_1, \ldots, x_k)$ and functions $a_1, \ldots, a_k \in C(m_n)$, the function $f(a_1(x), \ldots, a_k(x))$ belongs to $C(m_n)$ as well. For example, $a \in C(m_n)$ implies $\exp a \in C(m_n)$, and if a does not vanish anywhere, then $a^{-1} \in C(m_n)$ and $\ln |a| \in C(m_n)$.

We will use the above description to decompose a function from a class $C(m_n)$ into a sum of functions from the same class rapidly decreasing on the corresponding semi-axes.

Lemma 4.2. *Given a function $\gamma \in C(m_n)$, there exist functions $\gamma_\pm \in C(m_n)$ such that $\gamma = \gamma_+ + \gamma_-$ and the function γ_+ is analytic in the strip*

$$\Pi^+ = \{z = x + iy, \ x > \rho, \ | y | \leq 1\}$$

with some $\rho > 0$ and satisfies an estimate

$$|\gamma_+(z)| \leq Ce^{-x^2}, \qquad z \in \Pi^+.$$

Similarly, γ_- is analytic in the symmetric strip

$$\Pi^- = \{z = x + iy, \ x < -\rho, \ | y | \leq 1\}$$

and satisfies the same estimate in Π^-.

Proof. Let $\hat{\gamma}$ be a continuation of γ related to interval $[-1, 1]$ and possessing property *i)* from above. According to the Green–Cauchy theorem there exists a disk $D_\rho = \{ z : | z | \leq \rho \}$ such that

$$\hat{\gamma}(z) = \frac{1}{2\pi i} \int_{|z|=\rho} \frac{e^{-z^2+\zeta^2}\hat{\gamma}(\zeta)}{\zeta - z} d\zeta - \frac{1}{\pi} \int\int_{|z|\leq\rho} \frac{e^{-z^2+\zeta^2}\bar{\partial}\hat{\gamma}(\zeta)}{\zeta - z} d\xi d\eta,$$

where $\zeta = \xi + i\eta$ and

$$| \bar{\partial}\hat{\gamma}(\zeta) | \leq h(Q \mid \eta \mid)$$

with some constant Q.

Let us set

$$\gamma_{\pm}(z) = \frac{1}{2\pi i} \int\limits_{C^{\pm}} \frac{e^{-z^2+\zeta^2}\hat{\gamma}(\zeta)}{\zeta - z} d\zeta - \frac{1}{\pi} \int\int\limits_{D^{\pm}} \frac{e^{-z^2+\zeta^2}\overline{\partial}\hat{\gamma}(\zeta)}{\zeta - z} d\xi d\eta$$

where

$$C^{\pm} = \{ \zeta \in \mathbb{C} : |\zeta| = \rho, \quad \mp\mathrm{Re}\zeta \geq 0 \}$$

and

$$D^{\pm} = \{ \zeta \in \mathbb{C} : |\zeta| \leq \rho, \quad \mp\mathrm{Re}\zeta \geq 0 \}.$$

The functions $\gamma_+(z)$ and $\gamma_-(z)$ are defined and analytic in the complements $\mathbb{C} \setminus D^+$ and $\mathbb{C} \setminus D^-$, respectively, and according to the equality

$$\gamma(x) = \gamma_+(x) + \gamma_-(x), \quad |x| < \rho,$$

their restrictions to $\mathbb{R}$ may be extended as functions of the class $C(m_n)$. □

4.4.2 Equations with constant coefficients

Let us consider a linear difference equation

$$T\varphi(x) \equiv \sum_{i=1}^{q} a_i \varphi(x + \alpha_i) = \gamma(x), \quad \alpha_1 < \alpha_2 \cdots < \alpha_q, \tag{4.4.2}$$

with constant coefficients $a_i \in \mathbb{R}$. It is evident that the operator T is well-defined in every space $C(m_n)$.

Theorem 4.4. *Equation* (4.4.2) *has a solution* $\varphi \in C(m_n)$ *for every function* $\gamma \in C(m_n)$.

Proof. Given a function $\gamma \in C(m_n)$, it is sufficient to construct solutions $\varphi_{\pm} \in C(m_n)$ of equations

$$T\varphi_{\pm}(x) = \gamma_{\pm}(x) \tag{4.4.3}$$

where $\gamma_{\pm}$ are functions from the previous section.

Consider equation (4.4.3) with respect to $\varphi_+(x)$. If

$$\psi(x) = \varphi_+(x + \alpha_1),$$

then equation (4.4.3) takes on the form

$$\psi(x) = T_1\psi(x) + \frac{1}{a_1}\gamma_+(x) \tag{4.4.4}$$

where

$$T_1\psi(x) = \sum_{i=2}^{q} b_i \psi(x + \beta_i)$$

and

$$b_i = -\frac{a_i}{a_1}, \quad \beta_i = \alpha_i - \alpha_1, \quad 2 \le i \le q.$$

Note that

$$T_1^n \psi(x) = \sum_{2 \le i_1, \dots, i_n \le q} b_{i_1} b_{i_2} \cdots b_{i_n} \psi(x + \beta_{i_1} + \cdots + \beta_{i_n}).$$

Since $\alpha_i - \alpha_1 > 0$, $i \ge 2$, then for every bounded interval I there exist numbers $c = c(I)$ and $N = N(I)$ such that

$$x + \beta_{i_1} + \cdots + \beta_{i_n} > cn, \quad x \in I, \quad n \ge N. \tag{4.4.5}$$

Now let us set

$$\psi(x) = \frac{1}{a_1} \sum_{n=0}^{\infty} T_1^n \gamma_+(x)$$

and show that $\psi(x)$ belongs to $C(m_n)$. To this end we fix an interval I and choose a number N such that (4.4.5) holds. Then

$$\psi(x) = \psi_1(x) + \psi_2(x)$$

where

$$\psi_1(x) = \frac{1}{a_1} \sum_{n \le N} T_1^n \gamma_+(x)$$

and

$$\psi_2(x) = \frac{1}{a_1} \sum_{n > N} T_1^n \gamma_+(x).$$

The function $\psi_1(x)$, being a finite sum of functions from class $C(m_n)$, belongs to this class.

Furthermore we note that every function

$$h_n(x) = \frac{1}{a_1} T_1^n \gamma_+(x), \quad n > N,$$

may be continued as a complex analytic function in the strip $\prod^+$ and an estimate

$$| h_n(z) | \le c_1 e^{-c_2 n^2}, \quad n > N, \quad c_i = c_i(I),$$

easily follows from Lemma 4.2. Hence the function $\psi_2(x)$ is continued as a complex analytic function to the same strip $\prod^+$.

It is evident that the function $\psi(x) = \psi_1(x) + \psi_2(x)$ satisfies equation (4.4.4) and $\varphi_+(x) = \psi(x - \alpha_1)$ is a solution of (4.4.3) corresponding to γ_+.

A similar construction yields a solution of the equation

$$T\varphi_-(x) = \gamma_-(x)$$

belonging to the class $C(m_n)$. □

4.4.3 Equations with non-constant coefficients

Here we consider equation (4.4.1) with non-constant coefficients $a_i \in C(m_n)$.

Theorem 4.5. *Let*

$$a(x) \in C(m_n), \quad a(x) \neq 0, \quad x \in \mathbb{R}, \quad \alpha \neq 0.$$

Then the equation

$$\varphi(x) + a(x)\varphi(x + \alpha) = \gamma(x) \tag{4.4.6}$$

has a solution $\varphi \in C(m_n)$ *for any* $\gamma \in C(m_n)$.

Proof. Let $\psi, g \in C(m_n)$ be solutions of equations

$$g(x) - g(x + \alpha) = \ln |a(x)|$$

and

$$\psi(x) + \mathrm{sign}(a(x))\, \psi(x + \alpha) = \gamma(x)e^{-g(x)}$$

which exist according to Theorem 4.4. Then $\varphi(x) = e^{g(x)}\psi(x)$ is a solution of (4.4.6) belonging to $C(m_n)$. □

Condition $a(x) \neq 0$ may be replaced by requiring that the function $a(x)$ in (4.4.6) has finitely many zeros. On the other hand, if $a(x)$ vanishes on an infinite set, the solvability may fail.

Example 4.9. Let us show that the equation

$$\varphi(x) - (\sin 2\pi x)\varphi(x + 1) = 1 \tag{4.4.7}$$

has no quasianalytic solutions. Indeed, if $\varphi \in C(m_n)$ is such a solution, then the function

$$\psi(x) = (1 - \sin\, 2\pi x)\varphi(x) - 1$$

belongs to $C(m_n)$ and satisfies the homogeneous equation

$$\psi(x) = (\sin 2\pi x)\psi(x + 1).$$

Therefore

$$\psi(x) = (\sin 2\pi x)^p \psi(x + p), \quad p = 0, 1, \dots,$$

and all its derivatives at $x = 0$ vanish. If the class $C(m_n)$ is quasianalytic, then $\psi(x) \equiv 0$ which is impossible.

We will show later in Example 5.8, Section 5.2 that the operator

$$T\varphi(x) = \varphi(x) - (\sin 2\pi\theta x)\varphi(x + 1)$$

with a real rational number θ is surjective in C^∞ and therefore equation (4.4.7) has a C^∞-solution. On the other hand, if θ is irrational, then there exists an analytic function γ such that the equation $T\varphi(x) = \gamma(x)$ has no continuous solutions.

The decomposition method used to prove the global solvability of equation (4.4.2) with constant coefficients may be extended to equations with non-constant coefficients of a moderate growth at infinity, see [31]. We will show that the same method is effective without any restrictions on non-constant coefficients of equation (4.4.1) in non-quasianalytic classes $C(m_n)$. For this purpose we need functions from $C(m_n)$ vanishing on a ray.

Lemma 4.3. *In every non-quasianalytic class $C(m_n)$ there exists a function τ such that $0 \leq \tau(x) \leq 1$ and*

$$\tau(x) = \begin{cases} 0, & x \leq 0, \\ 1, & x \geq 1. \end{cases}$$

Proof. Since $C(m_n)$ is a non-quasianalytic class, there exists a function $h \in C(m_n)$ not equal identically to zero and satisfying $h^{(j)}(0) = 0, j = 0, 1, \ldots$. We can assume that $h(2^{-1}) = 1$. The function

$$g(x) = \begin{cases} 0, & x \leq 0, \\ h^2(x), & x \geq 0 \end{cases}$$

belongs to $C(m_n)$ and $g(2^{-1}) = 1$. The function $f(x) = g(x)g(1-x)$ belongs to $C(m_n)$ as well and $f(2^{-1}) = 1$. Since $f(x) = 0$ outside the interval $(0,1)$, the integral

$$c = \int_{-\infty}^{\infty} f(t)dt$$

converges and the function

$$\tau(x) = \frac{1}{c} \int_{-\infty}^{x} f(t)dt$$

satisfies all requirements of Lemma 4.3. □

Theorem 4.6. *Let $C(m_n)$ be a non-quasianalytic class and*

$$a_i \in C(m_n), \quad i = 1, \ldots, q; \quad a_1(x) \neq 0, \quad a_q(x) \neq 0, \quad x \in \mathbb{R}.$$

Then equation (4.4.1) has a solution $\varphi \in C(m_n)$ for every $\gamma \in C(m_n)$.

Proof. Let τ be a function from Lemma 4.3. It is sufficient to find solutions $\varphi_\pm \in C(m_n)$ of equations (4.4.3) where the functions

$$\gamma_-(x) = \tau(x)\gamma(x), \qquad \gamma_+(x) = (1 - \tau(x))\gamma(x)$$

belong to $C(m_n)$. In order to construct φ_+ we introduce, following the proof of Theorem 4.4, the function

$$\psi(x) = \varphi_+(x + \alpha_1)$$

and write the corresponding equation in the form

$$\psi(x) = T_1\psi(x) + \frac{\gamma_+(x)}{a_1(x)}$$

where

$$T_1\psi(x) = \sum_{i=2}^{q} b_i(x)\psi(x + \beta_i)$$

and

$$b_i(x) = -\frac{a_i(x)}{a_1(x)}, \quad \beta_i = \alpha_i - \alpha_1, \quad 2 \le i \le q.$$

Since $a_1(x) \neq 0$, the coefficients $b_i, i = 2, \dots, q$, and the function $\tilde{\gamma}_+ = \gamma_+/a_1$ belong to $C(m_n)$. For every bounded interval I of the real axis the series

$$\psi(x) = \sum_{n=0}^{\infty} T_1^n \tilde{\gamma}_+(x), \qquad x \in I,$$

contains finitely many non-vanishing terms and the function $\varphi_+(x) = \psi(x - \alpha_1)$ is a solution of (4.4.3) belonging to $C(m_n)$. A similar construction with $\psi(x) = \varphi_-(x + \alpha_q)$ yields its solution corresponding to $\gamma_-(x)$. □

4.5 Comments and references

4.1. Theorem 4.1 is a version of local theorems described in Chapter 3.
4.2. The notion of an absorber was first introduced in [19]. The decomposition method was first described in [31]. It is applicable to a study of multidimensional functional and differential-functional equations on finite-dimensional manifolds and even on Banach spaces [63]. In particular, equations

$$g(x, D_1\varphi(F_1(x)), \dots, D_n\varphi(F_n(x))) = 0$$

with differential operators

$$D_j = \sum_{k=0}^{p_j} a_{jk}\frac{d^k}{dx^k}$$

were studied in [28]. In [20] the same method was applied to systems of functional equations, see also [27].
4.3. It is evident that if the equation

$$g(x, \varphi(F(x))) = 0$$

with a C^k-diffeomorphism F is invertible with respect to y, then it has a global C^k-solution. It follows from results of Chapters 3 and 4 that every equation

$$g(x, \varphi(F_1(x)), \varphi(F_2(x))) = 0$$

with given C^k-diffeomorphisms F_1 and F_2 and arbitrary C^k-function $g(x, y_1, y_2)$, invertible with respect to y_1 and y_2, is globally C^k-solvable if and only if $F_1 F_2^{-1}$ has no fixed points. The natural question arises:

What are conditions on C^k-diffeomorphisms $F_1, \ldots, F_n$ such that every equation (4.0.1) *invertible with respect to variables $y_1, \ldots, y_n$ is globally C^k-solvable?*

Corollary 4.1 states some sufficient conditions, while Example 4.8 shows that they are not necessary even for affine transformations $F_j(x) = \lambda_j x + \mu_j$.

4.4. Theorem 4.4 was proved in [22]. The local analytic solvability of difference equation (4.4.1) in $\mathbb{R}^n, n \geq 1$, with arbitrary non-constant analytic coefficients was proved in [45]. A local solution on $\mathbb{R}^1$ is extended step-by-step to a global solution on the entire real line, but not on $\mathbb{R}^n$ with $n > 1$. The global solvability of multi-dimensional equation (4.4.1) in real-analytic functions was studied in [31].

Lemma 4.3 is a version of a result due to S.Mandelbroit [60].

As we have already noted, the decomposition method may be used to prove the solvability of equation (4.4.1) with non-constant coefficients in quasianalytic classes $C(m_n)$. For example, it is sufficient to assume that the functions

$$a_2(x)/a_1(x), \ \ldots, \quad a_q(x)/a_1(x), \quad a_1(x)/a_q(x), \quad \ldots, \quad a_{q-1}(x)/a_q(x)$$

may be continued in some strip $\{z = x + iy, |y| < c\}$ as analytic functions increasing not too rapidly at infinity. This condition holds if $a_1(x), \ldots, a_q(x)$ are rational functions.

The problem of solvability of equation (4.4.1) with arbitrary coefficients in quasianalytic classes remains unsolved.

Chapter 5

Linear Equations

In this chapter we study linear equations

$$L\varphi(x) = \gamma(x), \quad L\varphi(x) \equiv \sum_{j=1}^{n} a_j(x)\varphi(F_j(x)) \tag{5.0.1}$$

in a framework of a general theory of linear operators in linear topological spaces. Unlike in the previous chapters, here we assume the coefficients a_j and the function γ to be complex-valued functions and look for complex-valued solutions φ.

More exactly, given C^k-functions $a_j : M \to \mathbb{C}$, the operator L is well defined in spaces $C^p(M, \mathbb{C})$, $p = 0, 1, \dots, k$. We endow $C^p(M, \mathbb{C})$ with the topology of uniform convergence on compact subsets of all derivatives up to the order k. The space of analytic functions is not complete with respect to this topology and we consider here the case $k \neq \aleph$ only.

Since operator $L : C^p(M, \mathbb{C}) \to C^p(M, \mathbb{C})$ is continuous for $p = 0, 1, \dots, k$, its kernel $\operatorname{Ker} L \subset C^p(M, \mathbb{C})$ is a closed linear manifold, or *subspace*. In contrast to this, the image $\operatorname{Im} L$ is a linear manifold, generally speaking, non-closed.

Example 5.1. Given a C^k-function $a(x) \neq 0$, $x \in M$, the image of the *multiplication operator*

$$L\varphi(x) = a(x)\varphi(x), \quad x \in M,$$

coincides with $C^k(M, \mathbb{C})$. This statement fails if $a(x)$ vanishes at some points. For instance, the image of the operator L_0 of multiplication by $a(x) = x$ is closed in the space $C^\infty(\mathbb{R})$, but not in $C^k(\mathbb{R})$ with $k < \infty$. Indeed, if $\gamma \in \operatorname{Im} L_0$, then $\gamma(0) = 0$ and the function $\varphi(x) = x^{-1}\gamma(x)$ belongs to C^∞ as well. For every $k < \infty$ the function

$$\gamma(x) = \frac{x^k}{1 + |\ln|x||}, \quad x \neq 0; \ \gamma(0) = 0,$$

belongs to C^k, but $x^{-1}\gamma(x)$ does not. Hence $\gamma \notin \operatorname{Im} L_0$. On the other hand, every C^k-function γ, $\gamma(0) = 0$, may be approximated in $C^k(M, \mathbb{C})$ by functions $\gamma_n \in C^\infty$, $\gamma_n(0) = 0$, belonging to $\operatorname{Im} L_0$, and hence $\operatorname{Im} L_0$ is not a closed manifold.

In the previous example the number $\gamma(0)$ is a linear functional in the space of smooth functions and condition $\gamma(0) = 0$ is necessary for the solvability of the equation $L_0\varphi = \gamma$. For more general equations there may be more than one and even an infinite number of necessary conditions of such a type.

According to the Riesz Theorem every continuous linear functional $l \in (C^0(M, \mathbb{C}))^*$ has the form

$$l(\gamma) = \int \gamma(x) d\mu(x)$$

where μ is a complex-valued Borel measure with a compact support. If $k = 1, 2, \ldots, \infty$, then $l \in (C^k(M, \mathbb{C}))^*$ is a finite linear combination of functionals

$$l_j(\gamma) = \int \gamma^{(j)}(x) d\mu_j(x), \quad j = 0, 1, \ldots, k,$$

with μ_j of the same type.

A *support* of a functional $l \in (C^k(M, \mathbb{C}))^*$ is the set of all points $x \in M$ such that for every neighborhood U of x there is a function $\gamma \in C^k(M, \mathbb{C})$ vanishing outside U and such that $l(\gamma) \neq 0$.

Given a continuous linear operator $L : C^k(M, \mathbb{C}) \to C^k(M, \mathbb{C})$, the *dual operator* $L^* : (C^k(M, \mathbb{C}))^* \to (C^k(M, \mathbb{C}))^*$ is defined by the relation

$$L^* l(\gamma) = l(L\gamma).$$

If $\gamma \in \operatorname{Im} L$ and $L^* l = 0$, then $l(\gamma) = 0$. In other words, $\operatorname{Im} L$ is orthogonal to $\operatorname{Ker} L^*$, which means that $\operatorname{Im} L \subset (\operatorname{Ker} L^*)^\perp$. Moreover, it is known [59] that

$$\overline{\operatorname{Im} L} = (\operatorname{Ker} L^*)^\perp, \tag{5.0.2}$$

and hence if the manifold $\operatorname{Im} L$ is closed, then it coincides with the subspace $(\operatorname{Ker} L^*)^\perp$. In such a situation the solvability conditions for (5.0.1) are "algebraic":

$$\gamma \in \operatorname{Im} L \iff l(\gamma) = 0, \quad \gamma \in \operatorname{Ker} L^*.$$

The latter is a generalization of a well-known theorem on a system of linear equations from finite-dimensional linear algebra.

An operator whose image is closed is called *normally solvable* and equation (5.0.1) with such an operator is also called normally solvable. Applying to L the theorem on an inverse homomorphism [59], it is easy to check that L is normally solvable if and only if for every sequence $\{\gamma_n\} \subset \operatorname{Im} L$, $\gamma_n \to 0$, there exists a sequence $\{\varphi_n\}, \varphi_n \to 0$, such that $L\varphi_n = \gamma_n$.

The space Ker L^* is called the *cokernel* of operator L and is denoted by Coker L.

A normally solvable operator L and the corresponding equation are called *Fredholm* if both subspaces Ker L and Coker L are finite-dimensional. The number

$$\operatorname{ind} L = \dim \operatorname{Ker} L - \dim \operatorname{Coker} L$$

is called the *index* of L. The index is invariant with respect to small perturbations of L.

A normally solvable operator L and the corresponding equation are called *semi-Fredholm* if

$$\min(\dim \operatorname{Ker} L, \dim \operatorname{Coker} L) < \infty.$$

If $k < \infty$, then $C^k(\mathbb{T}, \mathbb{C})$ is a Banach space with the norm

$$\|\gamma\| = \max_{0 \le j \le k} \max_{x \in \mathbb{T}} |\gamma^{(j)}(x)|$$

which allows us to apply the spectral theory of linear operators for investigation of equation (5.0.1).

The *spectrum* of a continuous linear operator L in a Banach space $\mathcal{B}$ is the set $\operatorname{spec} L$ of all complex numbers λ such that the operator $L - \lambda I$ is not a bijective mapping of $\mathcal{B}$. It means that $\lambda \in$ spec L if and only if either Ker $(L - \alpha I) \neq \{0\}$ or Im $(L - \alpha I) \neq \mathcal{B}$. If $\lambda \notin \operatorname{spec} L$, then there exists the inverse linear operator $(L - \lambda I)^{-1}$ and, according to the Banach Theorem [59], it is continuous in $\mathcal{B}$. This definition of the spectrum is suitable for a wide range of linear topological spaces for which the Banach Theorem is valid.

The spectrum of an operator in a Banach space is a non-empty compact set and the number

$$r(L) = \max_{\lambda \in \operatorname{spec} L} |\lambda|$$

is called a *spectral radius* of L. According to this definition

$$\operatorname{spec} L \subset \{\, \lambda \in \mathbb{C} : |\lambda| \le r(L) \,\}.$$

The spectral radius may be calculated using the *Gel'fand formula* [59]

$$r(L) = \lim_{n \to \infty} \sqrt[n]{\|L^n\|}.$$

Example 5.2. Given a continuous function $b : \mathbb{T} \to \mathbb{C}$, consider the linear operator

$$L\varphi(z) = b(z)\varphi(\rho_\theta(z)), \quad z \in \mathbb{T},$$

with an irrational rotation $\rho_\theta : \mathbb{T} \to \mathbb{T}$. The space $C^0(\mathbb{T}, \mathbb{C})$ of continuous functions on $\mathbb{T}$ may be identified with the space of 1-periodic continuous functions on $\mathbb{R}$, and the operator L may be written in the form

$$L\varphi(x) = b(x)\varphi(x + \theta), \quad x \in \mathbb{R},$$

where b is a continuous 1-periodic function and θ is an irrational number. Let us find $\operatorname{spec} L$ under assumption $b(x) \neq 0,\ x \in \mathbb{R}$. To find first the spectral radius of L we note that

$$\|L^n\| = \max_{x\in\mathbb{R}} |b(x) \dots b(x + (n-1)\theta)|.$$

According to the Weyl theorem [68] the identity

$$\lim_{n\to\infty} \frac{1}{n}\sum_{j=0}^{n-1} \ln|b(x + j\theta)| = \int_0^1 \ln|b(t)|dt$$

holds uniformly with respect to x. Now the Gel'fand formula implies

$$r(L) = \exp\int_0^1 \ln|b(x)|dx.$$

Similarly, the spectral radius of the inverse operator

$$L^{-1}\varphi(x) = \frac{\varphi(x-\theta)}{b(x)}$$

is equal to

$$r(L^{-1}) = \exp\int_0^1 \ln\frac{1}{|b(x)|}dx = \frac{1}{r(L)}.$$

If $\lambda \in \operatorname{spec} L$, then $\lambda^{-1} \in \operatorname{spec} L^{-1}$. Hence $\operatorname{spec} L$ is contained in a circle of radius $r(L)$:

$$\lambda \in \operatorname{spec} L \implies |\lambda| = r(L).$$

Furthermore, $\operatorname{spec} L$ is invariant with respect to rotations of $\mathbb{T}$. Indeed, set

$$R_k\varphi(x) = e^{-2\pi ikx}\varphi(x).$$

Then R_k is a continuous operator in $C^0(\mathbb{R},\mathbb{C})$ and

$$R_kLR_k^{-1}\varphi(x) = e^{2\pi ik\theta}L\varphi(x).$$

Since $\operatorname{spec} R_kLR_k^{-1} = \operatorname{spec} L$, the implication

$$\lambda_0 \in \operatorname{spec} L \implies e^{2\pi ik\theta}\lambda_0 \in \operatorname{spec} L, \quad k = 0, 1, \dots,$$

is true, and since θ is irrational, the set $\operatorname{spec} L$ is dense in the circle $\{ z \in \mathbb{C} : |z| = r(L) \}$. Being closed, it coincides with this circle.

Thus the operator

$$L\varphi(x) - \lambda\varphi(x) \equiv b(x)\varphi(x + \theta) - \lambda\varphi(x)$$

is invertible in the space of continuous functions on $\mathbb{T}$ if and only if

$$|\lambda| \neq \exp\int_0^1 \ln|b(x)|dx.$$

Let us show that if $b(x) \equiv 1$ and hence

$$\operatorname{spec} L = \{\lambda \in \mathbb{C} : |\lambda| = 1\},$$

then $\operatorname{Im}(L - \lambda I)$ is not a closed manifold for $\lambda \in \operatorname{spec} L$.

If $\lambda = e^{2\pi i\alpha}$, $\alpha \in \mathbb{R}$, and φ is a solution of the equation

$$\varphi(x + \theta) - e^{2\pi i\alpha}\varphi(x) = \gamma(x),$$

then its Fourier coefficients

$$\varphi_n = \int_0^1 \varphi(x)e^{-2\pi inx}dx, \quad n \in \mathbb{Z},$$

satisfy the equalities

$$\left(e^{2\pi in\theta} - e^{2\pi i\alpha}\right)\varphi_n = \gamma_n, \quad n \in \mathbb{Z},$$

where

$$\gamma_n = \int_0^1 \gamma(x)e^{-2\pi inx}dx$$

are the Fourier coefficients of γ. Therefore if $n\theta - \alpha \notin \mathbb{Z}$ for all $n \in \mathbb{Z}$, then

$$\varphi_n = \frac{\gamma_n}{e^{2\pi in\theta} - e^{2\pi i\alpha}}.$$

In particular, every trigonometric polynomial

$$P(x) = \sum_{n=-N}^{N} P_n e^{inx}$$

belongs to $\operatorname{Im}(L - \lambda I)$. Hence

$$\overline{\operatorname{Im}(L - \lambda I)} = C^0(\mathbb{T}, \mathbb{C})$$

and $\operatorname{Coker}(L - \lambda I) = \{0\}$. If $n_0\theta - \alpha \in \mathbb{Z}$ for some $n_0 \in \mathbb{Z}$, then the number n_0 with this property is unique and a trigonometric polynomial P belongs to $\operatorname{Im}(L - \lambda I)$ if and only if $P_{n_0} = 0$. Hence in this case

$$\overline{\operatorname{Im}(L - \lambda I)} = \{ \gamma \in C^0(\mathbb{T}, \mathbb{C}) : \gamma_{n_0} = 0 \},$$

and $\dim \operatorname{Coker}(L - \lambda I) = 1$.

If we choose now a sequence of integers $n_j > |n_0|$, $n_j \to \infty$, such that

$$\sum_{j=1}^{\infty} |\varepsilon_j| < \infty, \qquad \varepsilon_j = e^{2\pi i n_j \theta} - e^{2\pi i \alpha},$$

then the series

$$\gamma(x) = \sum_{j=1}^{\infty} \varepsilon_j e^{2\pi i n_j x}$$

defines a continuous 1-periodic function. In any case it belongs to $\overline{\operatorname{Im}(L - \lambda I)}$ but does not belong to $\operatorname{Im}(L - \lambda I)$: otherwise there exists a continuous 1-periodic function

$$\varphi(x) = \sum_{n=1}^{\infty} \varphi_n e^{2\pi i n x}$$

with the Fourier coefficients

$$\varphi_n = 0, \quad n \neq n_j, \quad \varphi_{n_j} = \frac{\gamma_{n_j}}{e^{2\pi i n_j \theta} - e^{2\pi i \alpha}} = 1,$$

which is a contradiction to the Riemann–Lebesgue Theorem [59]. Therefore $\operatorname{Im}(L - \lambda I)$ is not a closed manifold for every $\lambda \in \mathbb{T}$.

All points of the spectrum of the operator L in Example 5.2 are not isolated, and they lie on its boundary. It is known [74] that an operator $L - \lambda I$ is not normally solvable for every such point λ belonging to the spectrum of L.

5.1 Generalized linear Abel equation

In this section we study equations

$$L\varphi(x) \equiv \varphi(F(x)) - a(x)\varphi(x) = \gamma(x), \quad x \in M, \tag{5.1.1}$$

from a point of view of the theory of linear operators. We assume that $F : M \to M$ is a C^k-mapping and a is a complex-valued C^k–function.

An inspection of Chapter 3 reveals that its methods and results related to linear equations may be extended (with obvious modifications) to similar equations with complex-valued functions. For example, if $F : \mathbb{R} \to \mathbb{R}$ is a fixed-point free

C^k-diffeomorphism and a is a non-vanishing complex-valued C^k-function, then equation (5.1.1) may be reduced to the form

$$\psi(t + 1) = \psi(t)$$

using a transformation

$$\Phi_j(x,y) = (H(x),\ S(x)y + h(x))$$

with a C^k-diffeomorphism $H : \mathbb{R} \to \mathbb{R}$, a non-vanishing $C^k(\mathbb{R},\mathbb{C})$-function S and $h \in C^k(\mathbb{R},\mathbb{C})$. As a result L is a surjective operator, Coker $L = \{0\}$ and Ker L is an infinite-dimensional space.

To describe the general background of our method let, as in Section 3.3, $\{U_j\}$ be an open $F-$invariant covering of M. We denote by $Z(k,\{U_j\},L)$ the space of all cocycles with the topology induced by the topological product

$$Z^0(k,\{U_j\}) = \prod_j C^k(U_j,\mathbb{C}).$$

The set $B(k,\{U_j\},L)$ of all cocycles equivalent to zero is a linear subspace in $Z(k,\{U_j\},L)$, maybe not closed.

Proposition 5.1. *Assume that operator L of the form* (5.1.1) *is normally solvable in $C^k(U_j,\mathbb{C})$ for all j's. Then L is normally solvable in $C^k(M,\mathbb{C})$ if and only if the set $B(k,\{U_j\},L)$ is closed in $Z(k,\{U_j\},L)$.*

Proof. Let $B(k,\{U_j\},L)$ be closed, $\gamma_m \in \operatorname{Im} L$ and $\gamma_m \to \gamma$. Since $\gamma_m|_{U_j} \to \gamma|_{U_j}$ and L is normally solvable in $C^k(U_j,\mathbb{C})$, there exists a sequence $\{\varphi_{m,j}\} \in C^k(U_j,\mathbb{C})$ such that $L\varphi_{m,j}(x) = \gamma_m(x)$, $x \in U_j$, and $\varphi_{m,j} \to \varphi_j$ with $L\varphi_j(x) = \gamma(x)$, $x \in U_j$. If cocycles $r_m = \{r_{ij,m}\}$ and $r = \{r_{ij}\}$ are defined by

$$r_{ij,m}(x) = \varphi_{m,j}(x) - \varphi_{m,i}(x), \quad r_{ij}(x) = \varphi_j(x) - \varphi_i(x), \quad x \in U_i \bigcap U_j,$$

then $r_m \to r$. According to Theorem 3.10 the assumption $\gamma_m \in \operatorname{Im} L$ implies that all cocycles r_m are equivalent to zero and hence belong to $B(k,\{U_j\},L)$. Since the latter space is closed, $r \in B(k,\{U_j\},L)$ as well and again using Theorem 3.10 we conclude that $\gamma \in \operatorname{Im} L$.

Conversely, let us assume Im L to be closed. Following [44] introduce the space of 1-cocycles

$$Z^1(k,\{U_j\}) \subset \prod_{j,i} C^k(U_i \cap U_j,\mathbb{C})$$

formed by collections

$$r = \{r_{ij}\}, \qquad r_{ij} \in C^k(U_i \cap U_j,\mathbb{C}),$$

satisfying the conditions

$$r_{ij} = -r_{ji}, \quad r_{ij}(x) + r_{jk}(x) + r_{ki}(x) = 0, \quad x \in U_i \cap U_j \cap U_k.$$

It is known that the coboundary operator

$$\delta : Z^0(k, \{U_j\}) \to Z^1(k, \{U_j\})$$

defined by

$$(\delta(\{\varphi_k\}))_{ij}(x) = \{\varphi_j(x) - \varphi_i(x)\}, \quad x \in U_i \cap U_j,$$

is surjective [44, 46].

Let now a sequence $r_m = \{r_{ij,m}\} \in B(k, \{U_j\}, L)$ converge to $r = \{r_{ij}\} \in Z(k, \{U_j\}, L)$. Let us show that $r \in B(k, \{U_j\}, L)$.

Since δ is a surjective operator, there exist functions $\varphi_j \in C^k(U_j)$ such that $r_{ij}(x) = \varphi_i(x) - \varphi_j(x), x \in U_i \cap U_j$. Similarly, $r_{ij,m}(x) = \varphi_{i,m}(x) - \varphi_{j,m}(x), x \in U_i \cap U_j$ where $\varphi_{j,m} \in C^k(U_j)$. In addition, since δ is normally solvable, we can assume that $\varphi_{j,m}$ converges to φ_j in $C^k(U_j)$ for every j.

The local identities

$$\gamma(x) = L\varphi_j(x), \quad \gamma_m(x) = L\varphi_{j,m}(x), \quad x \in U_j,$$

define the the global functions γ and γ_m from the space $C^k(M, \mathbb{C})$ and $\gamma_m \to \gamma$. Since $r_m \in B(k, \{U_j\}, L)$, Theorem 3.10 implies $\gamma_m \in \operatorname{Im} L$. According to our assumption Im L is a closed subspace and therefore $\gamma \in \operatorname{Im} L$ as well. Again applying Theorem 3.10 we conclude $r \in B(k, \{U_j\}, L)$. □

5.1.1 Equations on the real line with a unique fixed point

Let us start with a homogeneous equation

$$L\varphi(x) \equiv \varphi(F(x)) - a(x)\varphi(x) = 0, \quad x \in \mathbb{R}. \tag{5.1.2}$$

We assume here that x_0 is a unique fixed point of the C^k-diffeomorphism $F : \mathbb{R} \to \mathbb{R}$ and $a \in C^k(\mathbb{R}, \mathbb{C})$, $a(x) \neq 0$, $x \in \mathbb{R}$.

If F preserves the orientation on $\mathbb{R}$, then according to Lemma 2.1 there exists a pair of straightening C^k-diffeomorphisms $\Psi_\pm(x)$ for F at x_0. Therefore every solution of (5.1.2) in a neighborhood of x_0 has the form

$$\varphi(x) = \begin{cases} h_+(\Psi_+(x)), & x > x_0, \\ h_-(\Psi_-(x)), & x < x_0, \end{cases}$$

where $h_\pm(t)$ are solutions of equations

$$h_\pm(t + 1) = a(\Psi_\pm^{-1}(t))h_\pm(t), \qquad t \in \mathbb{R}.$$

In the case $F'(x_0) < 0$ a general solution of (5.1.2) has the form

$$\varphi(x) = \begin{cases} h(x), & x > x_0, \\ a(F^{-1}(x))h(F^{-1}(x)), & x < x_0, \end{cases}$$

where h is a solution of the homogeneous equation

$$h(F^2(x)) = a(x)a(F(x))h(x).$$

This observation allows us to obtain an explicit description of all solutions in some particular cases.

Example 5.3. A general solution of the equation

$$L\varphi(x) \equiv \varphi(\lambda x) - \mu\varphi(x) = 0, \ \lambda \neq 1, \ \mu \in \mathbb{C}, \tag{5.1.3}$$

has the form

$$\varphi(x) = \begin{cases} x^\alpha \, \tau_+ \left(\dfrac{\ln x}{\ln \lambda}\right) e^{i\theta \ln x}, & x > 0, \\ |x|^\alpha \tau_- \left(\dfrac{\ln |x|}{\ln \lambda}\right) e^{i\theta \ln |x|}, & x < 0, \end{cases} \tag{5.1.4}$$

if $\lambda > 0$ and

$$\varphi(x) = \begin{cases} x^\alpha \, \tau \left(\dfrac{\ln x}{2\ln |\lambda|}\right) e^{i\theta \ln x}, & x > 0, \\ |x|^\alpha \tau \left(\dfrac{\ln |x\lambda^{-1}|}{2\ln |\lambda|}\right) e^{i\theta \ln |x|}, & x < 0, \end{cases} \tag{5.1.5}$$

if $\lambda < 0$. In both cases

$$\alpha = \frac{\ln |\mu|}{\ln |\lambda|}, \quad \theta = \frac{\arg \mu}{\ln |\lambda|},$$

and $\tau_\pm$, τ are arbitrary 1-periodic functions. Therefore the space of all C^k-solutions of (5.1.3) is infinite-dimensional if $k < \alpha$ and finite-dimensional otherwise.

Example 5.4. Consider an equation

$$\varphi(G_{r,\beta}(x)) - \mu\varphi(x) = 0, \quad \mu \in \mathbb{C},$$

where $G_{r,\beta}(x)$ is given by (2.3.10)–(2.3.11).

As we have seen in Section 2.3.2 the straightening diffeomorphisms have the form

$$W_\pm(x) = \frac{1}{r}x^{-r} + \frac{\beta}{r}\ln |x|, \quad x \in \mathbb{R}_\pm,$$

and therefore every solution may be represented in a neighborhood of $x = 0$ in the form

$$\varphi(x) = \begin{cases} \tau_+ \left(r^{-1}x^{-r} + \beta r^{-1}\ln x\right) e^{(\alpha+i\theta)\left(r^{-1}x^{-r}+\beta r^{-1}\ln x\right)}, & x > 0, \\ \tau_- \left(r^{-1}x^{-r} + \beta r^{-1}\ln |x|\right) e^{(\alpha+i\theta)\left(r^{-1}x^{-r}+\beta r^{-1}\ln |x|\right)}, & x < 0, \end{cases}$$

where

$$\alpha = \ln|\mu|, \quad \theta = \arg \mu,$$

and $\tau_{\pm}$ are arbitrary 1-periodic functions. If r is even and $|\mu| < 1$, then for every 1-periodic C^∞-function $\tau_{\pm}$ the function $\varphi(x)$ is continued as a C^∞-function flat at $x = 0$. Hence the space of smooth solutions is infinite-dimensional. If r is even and $|\mu| \geq 1$, then every continuous solution of the homogeneous equation is constant.

If r is odd and $|\mu| \neq 1$, then the space of smooth solutions is infinite-dimensional. Namely, if $|\mu| < 1$, then the function φ with $\tau_- = 0$, an arbitrary $\tau_+ \in C^\infty$ and $\varphi(0) = 0$ belongs to C^∞. If $|\mu| > 1$, then φ with $\tau_+ = 0$, $\tau_- \in C^\infty$ and $\varphi(0) = 0$ belongs to C^∞.

Let us now consider equation (5.1.1) with a unique fixed point on the real line.

Theorem 5.1. *Let $F : \mathbb{R} \to \mathbb{R}$ be a C^k-diffeomorphism with a unique fixed point x_0 and let a be a non-vanishing C^k-function. Assume that one of the following conditions holds:*

i) $k < \infty$, $|F'(x_0)|^k \neq |a(x_0)|$;

ii) $k = \infty$ and the function $F(F(x)) - x$ is non-flat at x_0;

iii) $k = \infty$ and the function $|a(x)| - 1$ is non-flat at x_0.

Then operator (5.1.1) *is semi-Fredholm in $C^k(\mathbb{R}, \mathbb{C})$ with a finite-dimensional subspace* Coker L.

Proof. It follows from Theorem 3.2 in the case $k < \infty$ and from Theorem 3.3 in the case $k = \infty$ that the formal C^k-solvability at x_0 implies the local one. According to Theorem 3.7 a local solution is extendable to a global solution. Hence $\gamma \in \operatorname{Im} L$ if and only if the equation is formally solvable at x_0. In turn the formal solvability means that a finite number of derivatives of γ at x_0 satisfy a finite set of linear equations. Since all derivatives at x_0 are continuous linear functionals on $C^k(\mathbb{R}, \mathbb{C})$, equation (5.1.1) is normally solvable. □

Examples 5.3 and 5.4 show that the space Ker L may be infinite-dimensional and hence, generally speaking, L is not a Fredholm operator.

According to Corollary 3.5 the operator (5.1.3) is a "local model" of general operators (5.1.1) with hyperbolic fixed points. The following proposition describes its properties.

Theorem 5.2. *Let an operator L be of the form* (5.1.3) *with $\lambda \neq 0$, $\lambda \neq \pm 1$ and let*

$$\alpha = \frac{\ln|\mu|}{\ln|\lambda|}.$$

Then

i) L is normally solvable in $C^k(\mathbb{R}, \mathbb{C})$ if and only if $k \neq \alpha$;

ii) The space Ker $L \subset C^k(\mathbb{R}, \mathbb{C})$ *is finite-dimensional if and only if $k \geq \alpha$;*

iii) The space Coker $L \subset (C^k(\mathbb{R},\mathbb{C}))^*$ *is finite-dimensional for every* k, $0 \leq k \leq \infty$.

Proof. We start by proving *i*). If $|\lambda|^k \neq |\mu|$, then by Theorem 5.1 the subspace Im L is closed. Conversely, if $|\lambda|^k = |\mu|$, and

$$\gamma_0(x) = \int_0^x \frac{(x-t)^{k-1}}{(k-1)!}\frac{\exp(i\theta \ln|t|)}{1+|\ln|t||}dt, \quad \theta = \frac{\arg\mu - k\arg\lambda}{\ln|\lambda|}, \tag{5.1.6}$$

then $\gamma_0 \in C^k(\mathbb{R},\mathbb{C})$, $\gamma_0^{(j)}(0) = 0, j = 0,\dots,k$, and equation $L\varphi = \gamma_0$ is formally C^k-solvable. If ψ is its local C^k-solution at $x = 0$, then

$$\psi^{(k)}(x) = \left(\frac{\lambda^k}{\mu}\right)^n \psi^{(k)}(\lambda^n x) - \frac{1}{\mu}\sum_{p=0}^{n-1}\left(\frac{\lambda^k}{\mu}\right)^p \gamma_0^{(k)}(\lambda^p x),$$

and if, to be definite, $|\lambda| < 1$ then the sum here is bounded as $n \to \infty$. However,

$$\left(\frac{\lambda^k}{\mu}\right)^p \gamma_0^{(k)}(\lambda^p x) = \frac{e^{i\theta \ln|x|}}{1+|p\ln|\lambda| + \ln|x||}$$

and the sum is not bounded, implying $\gamma_0 \notin$Im L.

On the other hand, every monomial x^p, $p > k$, belongs to Im L. Since the function γ_0 may be approximated in $C^k(\mathbb{R},\mathbb{C})$ by polynomials $Q(x) = \sum_{p>k} a_p x^p$, it belongs to $\overline{\text{Im}\, L}$. Therefore the manifold Im L is not closed.

The statement *ii*) is contained in Example 5.3.

It remains to prove that dim Coker $L < \infty$ for every $k = 0, 1, \dots$.

In the absence of resonances, $\mu \neq \lambda^s$, $s = 0, \dots, k$, the equation (5.1.3) is C^k-formally solvable for every $\gamma \in C^k$, and $\overline{\text{Im}\, L} = C^k(\mathbb{R},\mathbb{C})$, since all polynomials belong to Im L. Hence in this case Coker $L = \{0\}$. If $\mu = \lambda^s$ for some $0 \leq s \leq k$, then the criterion of the formal solvability is

$$\gamma^{(s)}(0) = 0$$

which yields dim Coker $L = 1$. □

Corollary 5.1. *Operator* (5.1.3) *is semi-Fredholm if and only if* $k \neq \alpha$ *and is Fredholm if and only if* $k > \alpha$.

5.1.2 Cohomological equation

We consider the cohomological equation

$$L\varphi(x) \equiv \varphi(F(x)) - \varphi(x) = \gamma(x), \quad x \in M, \tag{5.1.7}$$

where $F : M \to M$ is an orientation-preserving C^k-diffeomorphism with a finite number of fixed points. It is evident that if Fix$F \neq \emptyset$ then Ker$L$ consists of constant functions.

Let us start with the case of a unique fixed point.

Theorem 5.3. *Let* Fix $F = \{x_0\}$ *and let one of the following conditions hold:*
i) $M = \mathbb{R}$, $k \geq 1$ *and the point* x_0 *is hyperbolic for* F;
ii) $M = \mathbb{R}$, $k = \infty$ *and* $F(F(x)) - x$ *is not flat at* x_0;
iii) $M = \mathbb{T}$, $k = \infty$ *and* $F(x) - x$ *is not flat at* x_0.
Then operator (5.1.7) *is normally solvable. In the cases i) and ii) it is Fredholm; in the case iii) it is semi-Fredholm with dim Coker* $L = \infty$.

Proof. All statements related to conditions *i)* and *ii)* are contained in Theorem 5.1.

If $M = \mathbb{T}$, then by Theorem 3.3 the formal solvability at x_0 implies its local C^∞-solvability. The condition of formal solvability means that a finite number of derivatives of a given function $\gamma \in C^\infty(\mathbb{T})$ at x_0 satisfy a finite system of linear equations. Therefore if $\gamma_m \to \gamma$ and equation (5.1.7) is locally C^∞-solvable for all γ_m, then it is locally solvable for γ as well. According to Corollary 3.6 the inclusion $\gamma \in \operatorname{Im} L$ is equivalent to $\rho_\gamma(x;\varphi) = \text{const}$ where $\rho_\gamma(x;\varphi)$ is defined in Section 3.2.2 for a local solution φ of (5.1.7) at x_0.

Given $x, y \in \mathbb{T} \setminus \{x_0\}$, the difference

$$l_{x,y}(\gamma) = \rho_\gamma(x;\varphi) - \rho_\gamma(y;\varphi)$$

does not depend on a choice of φ and is a continuous linear functional on the closed subspace $\mathcal{E}$ of C^∞-functions γ for which equation (5.1.7) has a local C^∞-solution. By the Hahn–Banach Theorem, $l_{x,y}(\gamma)$ is continued as a linear functional $\tilde{l}_{x,y} \in (C^\infty(\mathbb{T}))^*$. Now the condition $\gamma \in \operatorname{Im} L$ is equivalent to

$$\gamma \in \mathcal{E}, \quad \tilde{l}_{x,y}(\gamma) = 0, \; x, y \in \mathbb{T} \setminus \{x_0\},$$

and hence $\operatorname{Im} L$ is a closed subspace. Since the subspace Coker L contains $\tilde{l}_{x,y}$ for every $x, y \in \mathbb{T} \setminus \{x_0\}$, its dimension is infinite. □

Theorem 5.4. *Let* $F : M \to M$ *be an orientation-preserving* C^k*-diffeomorphism, Fix* $F = \{x_1, \dots, x_n\}$, $n \geq 2$, *and let* $\{U_j\}_{j=1}^n, x_j \in U_j$, *be the* F*-invariant covering of* M *defined at the beginning of Section* 3.3. *If equation* (5.1.7) *is normally solvable in every space* $C^k(U_j, \mathbb{C})$, $j = 1, \dots, n$, *then* L *is semi-Fredholm in* $C^k(M, \mathbb{C})$ *with* dim Coker $L = \infty$.

Proof. Let us remind the reader that if $\gamma \in C(k, \{U_j\}, \mathbb{C})$, then C^k-solutions on U_j are determined uniquely up to constants and are given by formulas (3.3.12)–(3.3.13). The space $B(k, \{U_j\}, L)$ for the cohomological equation consists of cocycles $r = \{r_{i,i+1}\}$ with

$$r_{i,i+1}(x) = c_i, \quad c_i = \text{const}, \quad x \in U_i \cap U_{i+1},$$

and hence it is closed. According to Proposition 5.1 the space Im L is closed as well.

Furthermore, for every $x, y \in U_i \cap U_{j+1}$ the expression

$$l_{x,y}(\gamma) = \sum_{s=-\infty}^{\infty} (\gamma(F^s(x)) - \gamma(F^s(y))$$

is a continuous linear functional on the subspace $C(k, \{U_j\}, L) \subset C^k(M, \mathbb{C})$. According to Theorem 3.11 a function $\gamma \in C(k, \{U_j\}, L)$ belongs to $\operatorname{Im} L$ if and only if

$$l_{x,y}(\gamma) = 0, \quad x, y \in (x_j, x_{j+1}).$$

Using the Hahn–Banach Theorem we can extend $l_{x,y}$ to a continuous linear functional $\tilde{l}_{x,y} \in (C^k(M, \mathbb{C}))^*$. Every such continuation belongs to Coker L. Since $x, y \in (x_j, x_{j+1})$ are arbitrary, it follows that dim Coker $L = \infty$. □

Theorems 5.3 and 5.4 leads us to the following statement.

Corollary 5.2. *Let $F : M \to M$ be an orientation-preserving C^k-diffeomorphism,* $\operatorname{Fix} F = (x_1, \ldots, x_n)$, $n \geq 2$. *Assume that either $1 \leq k < \infty$ and all points x_j are hyperbolic for F or $k = \infty$ and $F(x) - x$ is not a flat function at every point $x_j \in$* Fix F. *Then operator* (5.1.7) *is semi-Fredholm,* dim Coker $L = \infty$ *and* dim Ker $L = 1$.

5.1.3 Spectrum of a weighted shift operator

An operator of the form

$$T\varphi(x) = b(x)\varphi(F(x)) \tag{5.1.8}$$

is called the *weighted shift operator* [2]. If $F : \mathbb{T} \to \mathbb{T}$ is a C^k-diffeomorphism, then the operator $T\varphi(x) = \varphi(F(x))$ is an automorphism of the Banach algebra $C^k(\mathbb{T}, \mathbb{C})$. According to the well-known Kamovitz–Sheinberg theorem [50], if such an automorphism is non-periodic, then the spectrum of T contains the unit circle. If $|\sigma| \neq 1$, then the operator

$$L_\sigma\varphi(x) = \varphi(F(x)) - \sigma\varphi(x) \tag{5.1.9}$$

is invertible in $C^0(\mathbb{T}, \mathbb{C})$ and hence in this case spec$T = \mathbb{T}$. Since every point $\sigma \in \mathbb{T}$ is a non-isolated boundary point of spec $\mathbb{T}$, the operator $L_\sigma : C^0(\mathbb{T}, \mathbb{C}) \to C^0(\mathbb{T}, \mathbb{C})$ is not normally solvable for $\sigma \in \mathbb{T}$, see [74]. In particular, the cohomology equation is not normally solvable in $C^0(\mathbb{T}, \mathbb{C})$ if F is not a periodic homeomorphism.

If $F : \mathbb{T} \to \mathbb{T}$ is a C^k-diffeomorphism, $k < \infty$, $\operatorname{Fix} F \neq \emptyset$ and

$$|\, F'(x) \,| = 1, \quad x \in \operatorname{Fix} F,$$

then

$$\operatorname{spec} T = \left\{ \sigma \in \mathbb{C} : \min_{x \in \operatorname{Fix} F} |b(x)| \leq |\sigma| \leq \max_{x \in \operatorname{Fix} F} |b(x)| \right\}$$

for every non-vanishing function $b \in C^k(\mathbb{T},\mathbb{C})$, see [52, 2]. In particular, if $\operatorname{Fix} F = \{x_0\}$, then $\operatorname{spec} T$ is the circle

$$\{\, \sigma \in \mathbb{C} : |\sigma| = |b(x_0)| \,\}.$$

We conclude that equation (5.1.1) with a unique fixed point x_0 on $\mathbb{T}$ has a unique global C^k-solution for every $\gamma \in C^k(\mathbb{T},\mathbb{C})$ if and only if $|b(x_0)| \neq 1$ which completely agrees with Theorem 3.9. If $|b(x_0)| = 1$, then every $\sigma \in \mathbb{C}$, $|\sigma| = 1$, is a non-isolated boundary point of $\operatorname{spec} T$ and hence the operator (5.1.9) is not normally solvable in $C^k(\mathbb{T},\mathbb{C})$, in contrast with the case $k = \infty$ (see Theorem 5.3).

Theorem 5.5. *Let $F : \mathbb{T} \to \mathbb{T}$ be an orientation- preserving C^k-diffeomorphism, $1 \le k < \infty$, let $\operatorname{Fix} F = \{\, x_1, \ldots, x_n \,\}$ where all fixed points are hyperbolic and ordered according to the orientation of $\mathbb{T}$ and let F be locally C^k-conjugate with its linear part at x_j for every $j = 1, \ldots, n$.*

Given a non-vanishing complex-valued function $b \in C^k(\mathbb{T},\mathbb{C})$, set

$$r = \min_{1\le j\le n} \lambda_j^k |b(x_j)|, \qquad R = \max_{1\le j\le n} \lambda_j^k |b(x_j)|$$

where $\lambda_j = F'(x_j)$.

Then the spectrum of the operator (5.1.8) in $C^k(\mathbb{T},\mathbb{C})$ coincides with the set

$$S = \{\sigma \in \mathbb{C} : r \le |\sigma| \le R\} \bigcup \left\{ \bigcup_{1\le j\le n;\; 0\le l\le k} \{\lambda_j^l b(x_j)\} \right\}.$$

Proof. Since all fixed points of F are hyperbolic, their number $n = 2m$ is even. For $\sigma \in \mathbb{C}$ the operator

$$L_\sigma \varphi(x) = \varphi(F(x)) - \frac{\sigma}{b(x)} \varphi(x)$$

is invertible in $C^k(\mathbb{T},\mathbb{C})$ if and only if $\operatorname{Im} L_\sigma = C^k(\mathbb{T},\mathbb{C})$ and $\operatorname{Ker} L_\sigma = \{0\}$. If $\sigma = \lambda_j^l b(x_j)$ for some non-negative integer $l \le k$ and some $j = 1, \ldots, n$, then there exists $\gamma \in C^k(\mathbb{T},\mathbb{C})$ such that the equation

$$L_\sigma \varphi(x) = \gamma(x) \tag{5.1.10}$$

is not formally C^k-solvable at x_j and hence $\gamma \notin \operatorname{Im} L_\sigma$. Therefore, all numbers

$$\lambda_j^l b(x_j), \quad l = 0, \ldots, k, \quad j = 1, \ldots, n,$$

belong to $\operatorname{spec} T$.

Now let us consider the closed subspace $\mathcal{B}_0 \subset C^k(\mathbb{T},\mathbb{C})$ of all C^k-functions φ such that

$$\varphi(x_j) = \cdots = \varphi^{(k)}(x_j) = 0, \quad j = 1, \ldots, n.$$

It is evident that $\mathcal{B}_0$ is L_σ-invariant. Let us show that

$$\{\sigma : r \leq |\sigma| \leq R\} \subset \operatorname{spec} T|_{\mathcal{B}_0}. \tag{5.1.11}$$

As before, we set

$$U_1 = (x_n, x_2),\ U_j = (x_{j-1}, x_{j+1}),\ 2 \leq j \leq n-1,\ U_n = (x_{n-1}, x_1).$$

According to Corollary 3.5 the restriction of (5.1.10) to U_j may be reduced to (3.2.7) with

$$\lambda = \lambda_j, \qquad \mu = \frac{\sigma}{b(x_j)}.$$

A reducing transformation is given by formulas (3.3.14)–(3.3.17).

Denote

$$\beta_j(\sigma) = \frac{\ln|\sigma| - \ln|b(x_j)|}{\ln \lambda_j}, \quad j = 1, \ldots, n.$$

Let $k = \beta_j(\sigma)$ for some j and the function γ_0 be defined by (5.1.6) with above values λ and μ. Denote by γ any function from the space $\mathcal{B}_0$ which coincides with

$$h(x) = S_j^{-1}(x)\gamma_0(\Psi_j(x))$$

in a small neighborhood of x_0. Since equation (3.2.7) with $\tilde{\gamma} = \gamma_0$ has no local C^k-solutions at $t = 0$, equation (5.1.10) has no C^k-solutions. Hence the circles

$$C_j = \{\sigma \in \mathbb{C} : |\sigma| = \lambda_j^k |b(x_j)|\}, \quad j = 1, \ldots, n,$$

belong to $\operatorname{spec} T$.

Further, if $\gamma \in \mathcal{B}_0$, then (5.1.10) is formally C^k-solvable at x_j, $j = 1, \ldots, n$. Therefore, if $k \neq \beta_j(\sigma)$, then $\gamma \in \mathcal{B}_0$ implies $\gamma \in C(k, \{U_j\}, L_\sigma)$.

According to Theorem 3.14 the operator L_σ is not surjective if $k \geq \max(\beta_j(\sigma), \beta_{j+1}(\sigma))$ for some index j. On the other hand, it follows from the representation (5.1.4) of solutions of the homogeneous equation, that if $k < \min(\beta_j(\sigma), \beta_{j+1}(\sigma))$, then $\operatorname{Ker} L_\sigma \neq \{0\}$. Therefore $\sigma \notin \operatorname{spec} T|_{\mathcal{B}_0}$ only in the case

$$\min(\beta_j(\sigma), \beta_{j+1}(\sigma)) < k < \max(\beta_j(\sigma), \beta_{j+1}(\sigma)), \tag{5.1.12}$$

where $j = 1, \ldots, n$ and $\beta_{n+1}(\sigma) = \beta_1(\sigma)$. Since all fixed points are hyperbolic, the signs of numbers $\ln \lambda_1, \ldots, \ln \lambda_n$ interchange. Now a straightforward calculation shows that for a fixed j, relations (5.1.12) are fulfilled if and only if σ lies outside the annulus bounded by the circles C_j and C_{j+1}. Therefore a point σ satisfies (5.1.12) if and only if it is outside the union of these annuli and (5.1.11) holds.

Assume now that $\sigma \notin S$. Then equation (5.1.10) satisfies all conditions of Theorem 3.14 and according to its statement $\operatorname{Im} L_\sigma = C^k(\mathbb{T}, \mathbb{C})$. In addition, inequalities (5.1.12) and representations (5.1.4)–(5.1.5) imply $\operatorname{Ker} L_\sigma = \{0\}$ and therefore $\sigma \notin \operatorname{spec} T$. □

Note that the circles

$$C_{\min} = \{\sigma : |\sigma| = r\}, \qquad C_{\max} = \{\sigma : |\sigma| = R\}$$

consist of non-isolated boundary points of spec T. Hence the operator L_σ is not normally solvable if $\sigma \in C_{\min} \bigcup C_{\max}$. Let us show that L_σ is not normally solvable if

$$\sigma \in C_j, \ j = 1, \dots, n.$$

Indeed, the subspace

$$\mathcal{B}_j = \{\varphi \in C^k(\mathbb{T}, \mathbb{C}) : \varphi(x) = 0, \ x \notin [x_j, x_{j+1}]\}$$

is T-invariant. The previous arguments show that the spectrum of $T|_{\mathcal{B}_j}$ coincides with the annulus bounded by the circles C_j and C_{j+1}. Therefore the restriction of L_σ to $\mathcal{B}_j$ is not normally solvable if $\sigma \in C_j \cup C_{j+1}$. It means that for every such σ there exists a function $\gamma \in \mathcal{B}_j$ with $\gamma \notin L_\sigma(\mathcal{B}_j)$, but $\gamma = \lim \gamma_n, \gamma_n \in L_\sigma(\mathcal{B}_j)$. It is evident that $\gamma \notin L_\sigma(\mathcal{B}_0)$ and hence the restriction of L_σ to $\mathcal{B}_0$ is not normally solvable as well.

Now we note that

$$C^k(\mathbb{T}, \mathbb{C}) = \mathcal{B}_0 + \mathcal{F}$$

where $\mathcal{F}$ is the space of all trigonometric polynomials of degree not greater than $(k+1)n$ and the manifold

$$\operatorname{Im} L_\sigma = L_\sigma(\mathcal{B}_0) + L_\sigma(\mathcal{F})$$

cannot be closed.

5.1.4 Normal solvability of equations with hyperbolic fixed points

Here we consider operator (5.1.1) with a C^k-diffeomorphism F and an arbitrary non-vanishing C^k-function a.

Theorem 5.6. *Let a C^k-diffeomorphism $F : M \to M$, $k \geq 1$, preserve the orientation on M, let all its fixed points $\{x_1, \dots, x_n\}$, $n \geq 2$, be hyperbolic and let it be locally C^k-conjugate with its linear part at x_j for every $j = 1, \dots, n$. Set*

$$\lambda_j = F'(x_j), \quad \mu_j = a(x_j), \quad \alpha_j = \frac{\ln|\mu_j|}{\ln \lambda_j}.$$

Then

i) Operator (5.1.1) is normally solvable in $C^k(M, \mathbb{C})$ if and only if $k \neq \alpha_j$, $j = 1, \dots, n$.

ii) Operator (5.1.1) *is semi-Fredholm in* $C^k(M,\mathbb{C})$ *with* $\dim \operatorname{Ker} L < \infty$ *if and only if*

$$\alpha_1 < k; \quad \min(\alpha_j, \alpha_{j+1}) < k, \quad j = 1, \dots, n-1; \quad \alpha_n < k,$$

for $M = \mathbb{R}$ *and*

$$\min(\alpha_1, \alpha_n) < k; \quad \min(\alpha_j, \alpha_{j+1}) < k, \quad j = 1, \dots, n-1,$$

for $M = \mathbb{T}$.

iii) Operator (5.1.1) *is semi-Fredholm in* $C^k(M,\mathbb{C})$ *with* $\dim \operatorname{Coker} L < \infty$ *if and only if*

$$k < \max(\alpha_j, \alpha_{j+1}), \quad j = 1, \dots, n-1,$$

for $M = \mathbb{R}$ *and with the additional condition*

$$k < \max(\alpha_1, \alpha_n)$$

for $M = \mathbb{T}$.

Proof. Let $\{U_j\}_{j=1}^n, x_j \in U_j$, be the F-invariant covering of M defined at the beginning of Section 3.3.

To prove statement *i)* we note that if $k \neq \alpha_j, j = 1, \dots, n$, then by Theorem 5.1 the operators $L : C^k(U_j, \mathbb{C}) \to C^k(U_j, \mathbb{C})$ are normally solvable.

According to Theorem 3.12 every cocycle $h \in Z(k, \{U_j\}, L)$ is equivalent to its modified cocycle $\hat{h}$. Denote by $\hat{B}$ the subspace of modified cocycles from $B(k, \{U_j\}, L)$. If $r_m \in B(k, \{U_j\}, L)$ and $r_m \to r$, then $\hat{r}_m \to \hat{r}$. Since according to the same theorem $\dim \hat{B} < \infty$, we obtain $\hat{r} \in \hat{B}$. Hence $B(k, \{U_j\}, L)$ is closed and Proposition 5.1 implies that $\operatorname{Im} L$ is a closed subspace.

Now let us assume $k = \alpha_j$ for some j. Following Section 5.1.3 we consider the Banach space

$$\mathcal{B}_j = \{\varphi \in C^k(M, \mathbb{C}) : \varphi(x) = 0, \quad x \notin U_j \cap U_{j+1}\}$$

and the operator (5.1.8) with $b(x) = a^{-1}(x)$. The condition $\lambda_j^k = |a(x_j)|$ implies that $\sigma = 1$ is a non-isolated boundary point of its spectrum. Therefore the restriction $L|_{\mathcal{B}_j}$ is not normally solvable. Repeating arguments used at the end of Section 5.1.3 we conclude that $\operatorname{Im} L$ is not closed in $C^k(M, \mathbb{C})$.

To prove *ii)* we again make use of transformations (3.3.14)–(3.3.15). Under assumption $k < \alpha_1$ we set

$$\varphi(x) = \begin{cases} S_1^{-1}(x)\psi(\Psi_1(x)), & x < x_1, \\ 0, & x \geq x_1, \end{cases}$$

where ψ is a C^k-solution of the homogeneous equation

$$\psi(\lambda_1 t) = \mu_1 \psi(t), \quad t \in \mathbb{R}.$$

Since $\varphi \in \operatorname{Ker} L$ and the space of all such functions is infinite- dimensional, we obtain $\dim \operatorname{Ker} L = \infty$. Further, if $2 \le j \le n-1$ and $\psi(\lambda_j t) = \mu_j \psi(t)$, $t \in \mathbb{R}$, then the function

$$\varphi(x) = \begin{cases} S_j^{-1}(x)\psi(\Psi_j(x)), & x \in (x_j, x_{j+1}), \\ 0, & x \notin (x_j, x_{j+1}), \end{cases}$$

is a C^k-solution of the homogeneous equation on M. It follows from (5.1.4) that every such solution belongs to C^k if $k < \min(\alpha_j, \alpha_{j+1})$. Hence in this case dim Ker $L = \infty$ as well. Finally, if $k < \alpha_n$, then the function

$$\varphi(x) = \begin{cases} S_n^{-1}(x)\psi(\Psi_n(x)), & x > x_n, \\ 0, & x \le x_n, \end{cases}$$

is a C^k-solution of the homogeneous equation if $\psi(\lambda_n t) = \mu_n \psi(t)$ on $\mathbb{R}$.

On the other hand, if all inequalities from *ii*) hold, then for every $j = 1, 2, \ldots, n$ either the space $\operatorname{Ker} L|_{U_j}$ or the space $\operatorname{Ker} L|_{U_{j+1}}$ has a dimension ≤ 1. The union of all intervals with such a property covers the real axis except a finite number of points from Fix F. Therefore $\dim \operatorname{Ker} L < \infty$. The case $M = \mathbb{T}$ may be treated in a similar way.

To prove *iii*) let $M = \mathbb{R}$ and $k < \max(\alpha_j, \alpha_{j+1})$ for all $j = 1, \ldots, n-1$. Then by Theorem 3.13 a C^k-function γ belongs to $\operatorname{Im} L$ if and only if $\gamma \in C(k, \{U_j\}, L)$. The latter inclusion is equivalent to the formal C^k-solvability at every fixed point x_j. In the absence of resonances

$$\lambda_j \ne \mu_j^s, \quad s = 0, 1, \ldots, k-1, \; j = 1, \ldots, n,$$

equation (5.1.1) is solvable for every $\gamma \in C^k(M, \mathbb{C})$ and hence Coker $L = \{0\}$. If $\alpha_j = \mu_j^s$ for some $j = 1, 2, \ldots, n$ and some $s = 0, 1, \ldots, k-1$ and, as in Section 3.3.2,

$$\tilde{\gamma}_j(t) = S_j(\Psi_j^{-1}(\lambda_j t))\gamma(\Psi_j^{-1}(t)),$$

then the condition $\tilde{\gamma}_j^{(s)}(0) = 0$ is necessary and sufficient for (5.1.1) to be formally solvable at x_j. Since there are no more than n functionals

$$l_{s,j}(\gamma) = \tilde{\gamma}_j^{(s)}(0),$$

we conclude that dim Coker $L \le n$.

Assume now that $k > \max(\alpha_j, \alpha_{j+1})$ for some $j = 1, \ldots, n-1$. If $\gamma \in C(k, \{U_i\}, L)$, then the equation (5.1.1) has C^k-solutions φ_j on U_j and φ_{j+1} on U_{j+1}. If, in addition, $\gamma \in \operatorname{Im} L$, then

$$\rho(x) = \varphi_{j+1}(x) - \varphi_j(x) = \psi_{j+1}(x) - \psi_j(x), \; x \in (x_j, x_{j+1}),$$

where ψ_j, ψ_{j+1} are C^k-solutions of the homogeneous equation on U_j and U_{j+1}, respectively. Since $k > \max(\alpha_j, \alpha_{j+1})$ and for each j a resonance relation $\mu_j = \lambda_j^s$

is possible for a unique s only, the space of such solutions is of dimension not exceeding 1. As a result the inclusion $\gamma \in \operatorname{Im} L$ implies that the function $\rho(x)$ belongs to a space of dimension not exceeding 2.

In the absence of resonances

$$\mu_j \neq \lambda_j^s, \quad \mu_{j+1} \neq \lambda_{j+1}^s, \quad s = 0, 1, \ldots, k,$$

solutions φ_{j+1} and φ_j are unique and for a fixed $x \in (x_j, x_{j+1})$ the expression

$$l_x(\gamma) = \varphi_{j+1}(x) - \varphi_j(x)$$

defines a continuous linear functional on the closed subspace $C(k, \{U_i\}, L) \subset C^k(M, \mathbb{C})$. Everyone of its continuations to a functional $(C^k(M, \mathbb{C}))^*$ belongs to Coker L and since x is an arbitrary point in $(x_j, x_{j+1},)$ we conclude that dim Coker $L = \infty$.

If $\mu_j = \lambda_j^s$ for some $s = 0, 1, \ldots, k$ and there is no resonance relations $\mu_{j+1} = \lambda_{j+1}^r$, $r = 0, 1, \ldots, k$, then a C^k-solution on U_{j+1} is unique and a C^k-solution of (5.1.1) on U_j is defined up to a multiple of the function $g_j(x)$ given by (3.3.23). Hence if $\gamma \in \operatorname{Im} L$, then

$$\rho(x) = c g_j(x), \quad x \in (x_j, x_{j+1}),$$

with $c \in \mathbb{C}$. To eliminate the constant c we note that

$$\rho(x) g_j(y) - \rho(y) g_j(x) = 0, \quad x, y \in (x_j, x_{j+1}),$$

and the expression

$$l_{x,y}(\gamma) = \frac{\rho(x)}{g_j(x)} - \frac{\rho(y)}{g_j(y)} \tag{5.1.13}$$

does not depend on a choice of a solution φ_j and is a continuous linear functional on $C(k, \{\{U_j\}, L)$. Its Hahn–Banach extension to a linear functional on $C^k(M, \mathbb{C})$ belongs to Coker L and since x and y are arbitrary, the latter subspace is infinite-dimensional.

Finally, let

$$\mu_j = \lambda_j^{s_1}, \; \mu_{j+1} = \lambda_{j+1}^{s_2}$$

for some s_1, s_2, $0 \leq s_1, s_2 \leq k$. Then both C^k-solutions φ_j and φ_{j+1} are not unique and if $\gamma \in \operatorname{Im} L$, then

$$\rho(x) = \varphi_{j+1}(x) - \varphi_j(x) = c_1 g_j(x) + c_2 g_{j+1}(x), \; x \in (x_j, x_{j+1}),$$

with $c_1, c_2 \in \mathbb{C}$. To eliminate them we note that if

$$\Delta(x, y) \equiv \det \begin{pmatrix} g_j(x) & g_{j+1}(x) \\ g_j(y) & g_{j+1}(y) \end{pmatrix} = 0$$

for every $x, y \in (x_j, x_{j+1})$, then $g_{j+1}(x) = cg_j(x)$ with a constant $c \in \mathbb{C}$ and (5.1.13) defines a continuous linear functional on $C^k(M,\mathbb{C})$. Otherwise the expression

$$l_{x,y,z}(\gamma) = \frac{v(x,y)}{\Delta(x,y)} - \frac{v(x,z)}{\Delta(x,z)}$$

with

$$v(x,y) = \det\begin{pmatrix} g_j(x) & \rho(x) \\ g_j(y) & \rho(y) \end{pmatrix}$$

does not depend on a choice of solutions φ_j and φ_{j+1} for every triplet $x, y, z \in (x_j, x_{j+1})$ such that $\Delta(x,y) \neq 0$, $\Delta(y,z) \neq 0$. Again we obtain an expression which defines a continuous linear functional on $C(k, \{U_j\}, L)$ and conclude that $\dim \operatorname{Coker} L = \infty$.

The case $M = \mathbb{T}$ may be treated in a similar way. □

Corollary 5.3. *Under assumptions of Theorem* 5.6 *the operator* (5.1.1) *is Fredholm in $C^k(\mathbb{R},\mathbb{C})$ if and only if the number of fixed points of F is odd and*

$$\max_{i=0,1,\ldots,m} \alpha_{2i+1} < k < \min_{i=1,\ldots,m} \alpha_{2i}, \qquad m = \frac{n-1}{2}.$$

Such an operator is Fredholm in $C^k(\mathbb{T},\mathbb{C})$ if and only if

$$\min(\alpha_j, \alpha_{j+1}) < k < \max(\alpha_j, \alpha_{j+1}), \quad j = 1, \ldots, n-1,$$

and

$$\min(\alpha_1, \alpha_n) < k < \max(\alpha_1, \alpha_n).$$

Let us note that it easily follows from Theorem 5.6 that if $k = \infty$, then the operator (5.1.1) is semi-Fredholm in $C^\infty(M,\mathbb{C})$ with $\dim \operatorname{Ker} L < \infty$ and $\dim \operatorname{Coker} L = \infty$.

Example 5.5. Consider the operator

$$L\varphi(x) \equiv \varphi(x^3 + \rho x) - a(x)\varphi(x), \quad \rho \in (0,1),$$

from Example 3.10. Applying Theorem 5.6 we find that the space $\operatorname{Im} L$ is closed if and only if

$$k \neq \frac{\ln|a(-\sqrt{1-\rho})|}{\ln(3-2\rho)}, \quad k \neq \frac{\ln|a(0)|}{\ln\rho}, \quad k \neq \frac{\ln|a(\sqrt{1-\rho})|}{\ln(3-2\rho)}.$$

If these conditions are fulfilled, then the space $\operatorname{Ker} L$ is finite-dimensional if and only if

$$k > \max\left(\frac{\ln|a(-\sqrt{1-\rho})|}{\ln(3-2\rho)}, \frac{\ln|a(\sqrt{1-\rho})|}{\ln(3-2\rho)}\right),$$

and the space Coker L is finite-dimensional if and only if

$$k < \max\left(\frac{\ln|a(-\sqrt{1-\rho})|}{\ln(3-2\rho)}, \frac{\ln|a(0)|}{\ln\rho}\right)$$

and

$$k < \max\left(\frac{\ln|a(0)|}{\ln\rho}, \frac{\ln|a(0)|}{\ln(3-2\rho)}\right).$$

The operator L is Fredholm if and only if

$$\frac{\ln|a(0)|}{\ln\rho} > k > \max\left(\frac{\ln|a(-\sqrt{1-\rho})|}{\ln(3-2\rho)}, \frac{\ln|a(\sqrt{1-\rho})|}{\ln(3-2\rho)}\right).$$

In particular, if $a(x) = \mu$ is constant, then a criterion for L to be Fredholm is reduced to the inequality $|\mu| < \rho^k$.

Note that according to Corollary 5.3 the operator

$$L\varphi(x) = \varphi(\rho x) - \mu\varphi(x)$$

with a unique fixed point is Fredholm if and only if the inverse inequality $|\mu| > \rho^k$ holds.

5.1.5 Equations with periodic points

Theorem 5.6 is not immediately applicable if a mapping $F : \mathbb{T} \to \mathbb{T}$ has only two fixed points and inverts the orientation of $\mathbb{T}$. Fortunately, the mapping F^2 preserves the orientation and has the same fixed points. Moreover, F may have periodic points of any period $p \geq 2$ which, of course, are fixed points of F^p. If such a point is hyperbolic for F^p, it is said to be a *hyperbolic periodic point* for F itself.

Here we study equation (5.1.1) with a diffeomorphism $F : M \to M$ such that some its power preserves the orientation and has a finite number of fixed points.

Let us start with an auxiliary proposition.

Lemma 5.1. *Let A and B be continuous linear operators in a linear topological space $\mathcal{E}$ and let* Ker B $\subset$Im A. *If the operator BA is normally solvable, then the same is true for A.*

Proof. Let $\gamma_m \to \gamma, \gamma_m \in$Im$A$. We have to prove that $\gamma \in$ Im A as well. Let $\varphi_m \in \mathcal{E}$ be such that $\gamma_m = A\varphi_m$. Then $BA\varphi_m = B\gamma_m$ and hence $B\gamma_m \in$Im BA. Since the latter subspace is closed,

$$B\gamma = \lim_{m\to\infty} B\gamma_m \in \text{Im}\,(BA)$$

and $B\gamma = BA\varphi$ with some $\varphi \in \mathcal{E}$. Now the relations

$$\gamma - A\varphi \in \text{Ker}B \subset \text{Im}\, A$$

imply $\gamma - A\varphi = A\psi$ for some $\psi \in \mathcal{E}$ and therefore $\gamma = A(\varphi - \psi)$. □

Corollary 5.4. *Let, in addition to conditions of Lemma* 5.1, *the operators A and B commute. If the operator AB is semi-Fredholm, then A is semi-Fredholm as well and* $\mathrm{Ker}A \subset \mathrm{Ker}(AB), \mathrm{Coker}A \subset \mathrm{Coker}(BA)$.

Given an operator $T : \mathcal{E} \to \mathcal{E}$ and an integer $p \geq 1$, let us set $A = I - T$, $B = I + T + \cdots + T^{p-1}$. Then $AB = BA = I - T^p$. It follows from the identity

$$(I - T)\sum_{j=1}^{p-1} jT^{p-j-1} = pI - \sum_{j=0}^{p-1} T^j = pI - B$$

that $\mathrm{Ker}B \subset \mathrm{Im}A$.

Corollary 5.5. *Let a continuous operator $T : \mathcal{E} \to \mathcal{E}$ be such that $I - T^p$ is a normally solvable operator for some $p \geq 1$. Then $I - T$ is normally solvable as well. If, in addition, $I - T^p$ is semi-Fredholm, then $I - T$ has the same property,* $\mathrm{Ker}(I - T) \subset \mathrm{Ker}(I - T^p)$ *and* $\mathrm{Coker}(I - T) \subset \mathrm{Coker}(I - T^p)$.

Theorem 5.7. *Let $F : M \to M$ be a C^k-diffeomorphism such that some its power F^p satisfies conditions of Theorem* 5.6 *and let a be a non-vanishing C^k-function on M. Denote*

$$\lambda_j = (F^p)'(x_j), \quad \alpha_j = \frac{\ln |a(x_j) \, \ldots \, a(F^{p-1}(x_j))|}{\ln \lambda_j}, \quad j = 1, \ldots, n,$$

and assume $k \neq \alpha_j$, $j = 1, \ldots, n$. Then

i) the operator (5.1.1) *is normally solvable in the space* $C^k(M, \mathbb{C})$;

ii) if conditions ii) of Theorem 5.6 *hold, then the operator* (5.1.1) *is semi-Fredholm with* $\dim \mathrm{Ker}\, L < \infty$;

iii) if conditions iii) of Theorem 5.6 *hold, then the operator* (5.1.1) *is semi-Fredholm with* $\dim \mathrm{Coker}\, L < \infty$;

iv) if conditions of Corollary 5.3 *hold, then the operator* (5.1.1) *is Fredholm.*

Proof. It is sufficient to prove the theorem for the operator $I - L_1$ with

$$L_1\varphi(x) = a^{-1}(x)\varphi(F(x)).$$

Since

$$L_1^p\varphi(x) = a^{-1}(x) \, \ldots \, a^{-1}(F^{p-1}(x))\varphi(F^p(x)),$$

the equation $(I - L_1^p)\varphi = \gamma$ is equivalent to

$$\varphi(F^p(x)) - a(x) \, \ldots \, a(F^{p-1}(x))\varphi(x) = \tilde{\gamma}(x)$$

and all assertions follow from Theorem 5.6 and Corollary 5.5. □

Example 5.6. Let

$$F(x) = -x^3 - \rho x, \quad \rho \in (0, 1), \; x \in \mathbb{R}.$$

The mapping

$$F^2(x) = (x^3 + \rho x)((x^3 + \rho x)^2 + \rho)$$

has three fixed points $x_1 = 0$, $x_2 = -x_3 = \sqrt{1 - \rho}$. Since

$$\lambda_1 = (F^2)'(0) = \rho^2 < 1, \; \lambda_2 = \lambda_3 = (F^2)'(x_2) = (F^2)'(x_3) = (3 - 2\rho)^2 > 1,$$

all fixed points are hyperbolic. Therefore F has a fixed point $x_1 = 0$ and two periodic points

$$x_2 = \sqrt{1 - \rho}, \quad x_3 = -\sqrt{1 - \rho}.$$

Let

$$L\varphi(x) = \varphi(-x^3 - \rho x) - a(x)\varphi(x)$$

with a non-vanishing C^k-function $a(x)$. It follows from Theorem 5.7 that if

$$\rho^k \neq |a(0)|, \; (3 - 2\rho)^{2k} \neq |a(\sqrt{1 - \rho})a(-\sqrt{1 - \rho})|,$$

then L is a normally solvable operator in $C^k(\mathbb{R}, \mathbb{C})$. If, in addition,

$$\rho^{2s} \neq a^2(0) \;, \; (3 - 2\rho)^{2s} \neq a(\sqrt{1 - \rho})a(-\sqrt{1 - \rho}) \;, \quad 0 \leq s \leq k,$$

and

$$k < \max\left(\frac{\ln|a(0)|}{\ln\rho}, \frac{\ln|a(\sqrt{1 - \rho})\, a(-\sqrt{1 - \rho})|}{2\ln(3 - 2\rho)}\right),$$

then L is surjective in $C^k(\mathbb{R}, \mathbb{C})$.

In particular, the operator L with a constant coefficient $a(x) = a$ is normally solvable in $C^k(\mathbb{R}, \mathbb{C})$ if

$$\rho^k \neq |a|, \; (3 - 2\rho)^k \neq |a|,$$

and is surjective if, in addition,

$$\rho^{2s} \neq a^2, \; (3 - 2\rho)^{2s} \neq a^2, \; 0 \leq s \leq k,$$

$$k < \max\left(\frac{\ln|a|}{\ln\rho}, \frac{\ln|a|}{\ln(3 - 2\rho)}\right).$$

It follows from Theorem 5.7 that if all non-wandering points of C^k-diffeomorphism $F : M \to M$ with $k \geq 1$ are hyperbolic periodic points, then the cohomological equation is normally solvable in $C^k(M, \mathbb{C})$. In a more specific case of a preperiodic mapping F, the cohomological equation is normally solvable without any restriction on smoothness.

Let us remind the reader that a mapping $F : M \to M$ is called *preperiodic of period* $p \geq 1$ and of *preperiod* $l \geq 0$ if $F^{p+l}(x) = F^l(x)$, $x \in M$. In the case $l = 0$ the mapping F is periodic of period p. If F is a preperiodic mapping, then the semi-group

$$\mathfrak{G}(F) = \{ F^s, \ s = 1, 2, \dots, p + l \}$$

is finite. It is easy to check that if a function φ satisfies the cohomological equation

$$\varphi(F(x)) - \varphi(x) = \gamma(x)$$

with a preperiodic mapping F, then

$$\sum_{k=0}^{p-1} \gamma(F^{k+l}(x)) = 0, \quad x \in M.$$

Conversely, if the latter holds, then the function

$$\varphi(x) = -\sum_{j=0}^{l-1} \gamma(F^j(x)) + \frac{1}{p} \sum_{j=1}^{p-1} j\gamma(F^{j+l}(x))$$

is a solution of the cohomological equation. It follows from these formulas that if a C^k-mapping $F : M \to M$ is preperiodic, then the cohomological equation is normally solvable in the space $C^k(M, \mathbb{C})$. This assertion is true for all values $k \geq 1$.

Example 5.7. The continuous mapping

$$F(x) = \begin{cases} 1, & x \leq -1, \\ -x, & -1 \leq x \leq 1, \\ -1, & x \geq 1, \end{cases}$$

is preperiodic with period $p = 2$ and of preperiod $l = 1$. The cohomological equation

$$\varphi(F(x)) - \varphi(x) = \gamma(x)$$

has a continuous solution if and only if $\gamma(F(x)) + \gamma(F^2(x)) = 0$, and if the latter condition is satisfied, then

$$\varphi(x) = -\gamma(x) + \frac{1}{2}\gamma(F^2(x))$$

is a solution.

5.2 Localization of obstacles to solvability

Given an operator (5.0.1), let us set

$$S(L) = \bigcup_{l \in \operatorname{Coker} L} \operatorname{supp} l.$$

Loosely speaking (5.0.2) means that the set $S(L)$ contains all obstructions to solvability of (5.0.1). More exactly, if $\gamma \in C^k$, $\gamma|_{S(L)} = 0$, then $\gamma \in \overline{\operatorname{Im} L}$. Conversely, if a set $S \subset M$ is such that every C^k-function vanishing on S belongs to $\overline{\operatorname{Im} L}$, then $S \supset S(L)$. It permits us to apply results from the previous chapter to describe $S(L)$ via joint dynamical behavior of mappings $\{ F_j \}$.

Let us start from operator (5.0.1) in the form

$$L\varphi(x) \equiv \varphi(x) - \sum_{j=1}^{n} b_j(x)\varphi(H_j(x)), \quad x \in \mathbb{R}, \tag{5.2.1}$$

with C^k-functions b_j and H_j.

Theorem 5.8. *Let* $H_j(x) > x$, $x \in \mathbb{R}$, $j = 1, \dots , n$. *Then* $\operatorname{Im} L$ *is a dense manifold in the space* $C^k(\mathbb{R}, \mathbb{C})$ *and hence* $\operatorname{Coker} L = \{0\}$.

Proof. For every $c \in \mathbb{R}$ the semi-axis $[c, +\infty)$ is an absorber for the family $\{H_j\}$. According to Theorem 4.2 a C^k-function γ vanishing on $[c, +\infty)$ belongs to $\operatorname{Im} L$. The set of all such functions with arbitrary $c \in \mathbb{R}$ is dense in $C^k(\mathbb{R}, \mathbb{C})$. □

It is evident that the statement of Theorem 5.8 is true if $H_j(x) < x$, $x \in \mathbb{R}$, $j = 1, \dots, n$. In particular, if a C^k-mapping $F : \mathbb{R} \to \mathbb{R}$ is fixed-point free, then the image of the operator

$$L\varphi(x) = \varphi(x) - b(x)\varphi(F(x))$$

is dense in $C^k(\mathbb{R}, \mathbb{C})$, but, generally speaking, L is not surjective. For instance, if there exists a compact subset not wandering for F (cf., Examples 2.1 and 2.2), then equation (5.0.1) has no continuous solutions for some continuous functions γ. Further, the operator $L\varphi(x) = \varphi(x) - \varphi(F_\alpha(x))$ from Example 2.3 is surjective in $C^0(\mathbb{R}, \mathbb{C})$, but non-surjective in $C^k(\mathbb{R}, \mathbb{C})$, $k \geq 1$, although $\operatorname{Im} L$ is dense in $C^k(\mathbb{R}, \mathbb{C})$ for every $k = 0, 1, \dots , \infty$.

The following example shows that the property of the manifold $\operatorname{Im} L$ to be closed depends not only on a joint dynamical behavior of mappings H_j, but also on coefficients $b_j(x)$.

Example 5.8. Let

$$L\varphi(x) \equiv \varphi(x) - q\varphi(x + 1) \sin \pi\theta x = \gamma(x) \tag{5.2.2}$$

with $q \in \mathbb{C}$, $q \neq 0$. Here $F(x) = x + 1$, and if $|q| < 1$, then the conditions of Theorem 4.1 are satisfied with $V = \mathbb{R}$. Nevertheless, the latter is not applicable because V is not a compact set. On the other hand, the coefficient $b(x) = q \sin \pi\theta x$ vanishes on the infinite set

$$Z = \left\{ x : x = \frac{p}{\theta},\ p = 0, \pm 1, \ldots, \right\}$$

which does not permit us to apply Theorem 4.3.

We will show that if θ is rational and hence its orbit

$$\mathcal{N} = \bigcup_{n=-\infty}^{\infty} F^n(Z) = \left\{ x : x = \frac{p}{\theta} + n,\ p, n = 0, \pm 1, \ldots, \right\}$$

is discrete, then $L : C^k(\mathbb{R}, \mathbb{C}) \to C^k(\mathbb{R}, \mathbb{C})$ is a surjective operator, while if θ is irrational and hence $\mathcal{N}$ is dense in $\mathbb{R}$, then there exist C^∞-functions γ such that the equation $L\varphi = \gamma$ has no continuous solutions. Since the manifold $\operatorname{Im} L$ is dense, it is not closed in the latter case in $C^k(\mathbb{R}, \mathbb{C})$ for all $0 \leq k \leq \infty$.

We start with a rational $\theta = ts^{-1}$, t and s being mutually prime integers. In this case we have

$$\mathcal{N} = \left\{ \frac{p}{t},\ p = 0, \pm 1, \pm 2, \ldots \right\}$$

and there is the unique representation

$$\frac{p}{t} = \frac{ms}{t} - r, \quad r = 0, 1, \ldots, s - 1,\ m = 0, \pm 1, \ldots.$$

We set

$$\varphi_0\left(\frac{ms}{t}\right) = \gamma\left(\frac{ms}{t}\right), \quad m = 0, \pm 1, \ldots,$$

and define a solution φ_0 of (5.2.2) on $\mathcal{N}$ recursively;

$$\varphi_0\left(\frac{ms}{t} - r\right) = (-1)^{m+1} q \varphi_0\left(\frac{ms}{t} - r + 1\right) \sin \frac{\pi r t}{s} + \gamma\left(\frac{ms}{t} - r\right)$$

for $0 < r < s$.

Differentiating (5.2.2) we obtain equations

$$\varphi^{(j)}(x) - q\varphi^{(j)}(x + 1) \sin \pi\theta x = \gamma_j(x), \qquad j = 0, 1, \ldots, k,$$

where

$$\gamma_j(x) = \gamma^{(j)}(x) + \sum_{i=0}^{j-1} b_{ij}(x) \varphi^{(i)}(x)$$

with analytic functions b_{ij}. We use these equations to define

$$\varphi_0^{(j)}\left(\frac{ms}{t} - r\right) = (-1)^{m+1} q \varphi_0^{(j)}\left(\frac{ms}{t} - r + 1\right) \sin \frac{\pi r t}{s} + \gamma_j\left(\frac{ms}{t} - r\right)$$

for $0 < r < s$.

To extend the solution φ_0 from the set $\mathcal{N}$ to $\mathbb{R}$, let $\varphi_1(x)$, $x \in \mathbb{R}$, be a C^∞-function satisfying

$$\varphi_1^{(j)}\left(\frac{p}{t}\right) = \varphi_0^{(j)}\left(\frac{p}{t}\right).$$

Substituting $\varphi(x) = \psi(x) + \varphi_1(x)$ in (5.2.2) we arrive at the equation

$$\psi(x) = q\psi(x+1)\sin\pi\theta x + \tilde{\gamma}(x)$$

where the C^k-function

$$\tilde{\gamma}(x) = \gamma(x) - \varphi_1(x) + q\varphi_1(x+1)\sin\pi\theta x$$

vanishes at points pt^{-1} with all its derivatives up to order k.

Assume now $k < \infty$ and construct a C^k-function γ_1 vanishing in a neighborhood of the set $\mathcal{N}$ and such that

$$|\tilde{\gamma}^{(j)}(x) - \gamma_1^{(j)}(x)| \leq ce^{-x^2}, \; x \in \mathbb{R}, \; j = 0, 1, \dots, k, \tag{5.2.3}$$

with a constant $c > 0$. To this end we set $h(x) = e^{x^4}\tilde{\gamma}(x)$ and note that $h^{(j)}(s) = 0, j = 0, 1, \dots, k$ for all points $s \in \mathcal{N}$. For every such point s let $I(s, \delta) = \{x : |x - s| \leq 2\delta\}$ be a segment not containing points of $\mathcal{N}$ different from s and such that

$$|h^{(j)}(x)| \leq \delta^{k-j}, \quad x \in I(s,\delta), \quad j = 0, 1, \dots, k.$$

Define the function

$$\tau_s(x) = \begin{cases} 1, & |x - s| \leq \delta, \\ (2 - |x - s|\delta^{-1})^k, & \delta \leq |x - s| \leq 2\delta, \end{cases}$$

and set

$$H(x) = \begin{cases} \tau_s(x)h(x), & x \in I(s,\delta), \\ 0, & x \notin \bigcup_{s\in\mathcal{N}} I(s,\delta), \end{cases}$$

and

$$\gamma_2(x) = H(x)e^{-x^4}.$$

According to this definition γ_2 coincides with $\tilde{\gamma}$ in a neighborhood of $\mathcal{N}$ and therefore the function $\gamma_1 = \tilde{\gamma} - \gamma_2$ vanishes in the same neighborhood. Since

$$|\tau_s^{(j)}(x)| \leq k!\delta^{-j}, \quad x \in I(s,\delta), \quad j = 0, 1, \dots, k,$$

we have

$$|H^{(j)}(x)| \leq 2^k k!, \quad x \in I(s,\delta), \quad j = 0, 1, \dots, k.$$

Therefore

$$|\gamma_2^{(j)}(x)| \leq ce^{-x^2}, \; x \in \mathbb{R}, \; j = 0, 1, \dots, k,$$

which proves (5.2.3).

It is easy to see now that the series

$$\varphi(x) = \sum_{n=0}^{\infty} q^n \sin \pi\theta x \ldots \sin \pi\theta(x + n)\gamma_2(x + n)$$

defines a C^k-solution of (5.2.2) with $\gamma = \gamma_2$, implying $\gamma_2 \in \operatorname{Im} L$. It remains to show that $\gamma_1 \in \operatorname{Im} L$ as well. Let $\tau : \mathbb{R} \to [0,1]$ be a C^∞-function vanishing in $[0, +\infty)$ and equal to 1 in $(-\infty, -1]$. Then the function $\tau(x)\gamma_1(x)$ belongs to $\operatorname{Im} L$, since $[0, +\infty)$ is an absorber for $F(x) = x + 1$. Further, consider the equation

$$\varphi(x) = \frac{\varphi(x - 1)}{q \sin(\pi\theta(x-1))} - \frac{(1 - \tau(x - 1))\gamma_1(x - 1)}{q \sin(\pi\theta(x-1))}.$$

The operator

$$L_1\varphi(x) = \frac{\varphi(x - 1)}{q \sin(\pi\theta(x-1))}$$

is well defined in the space of C^k-functions vanishing in a neighborhood of $\mathcal{N}$. Since the function

$$g(x) = \frac{(1 - \tau(x - 1))\gamma_1(x - 1)}{q \sin(\pi\theta(x-1))}$$

belongs to this space and vanishes on the absorber $(-\infty, -1]$ of the inverse shift $x \to x - 1$, the function

$$\varphi(x) = \sum_{n=0}^{\infty} L_1^n g(x)$$

is a solution of the equation

$$\varphi(x) = L_1\varphi(x) + g(x).$$

Therefore the function $(1 - \tau(x))\gamma_1(x)$ belongs to $\operatorname{Im} L$ together with $\tau(x)\gamma_1(x)$ which proves the surjectivity of the operator L in $C^k(\mathbb{R}, \mathbb{C})$ for $k < \infty$.

If $k = \infty$, then the function $\tilde{\gamma}$ is flat at $\mathcal{N}$. The operator L_1 is well defined in the space of all such functions and we may repeat the previous arguments with $\gamma_2(x) \equiv 0$.

Let now θ be an irrational number. Denote by γ an entire analytic function such that

$$\left|\gamma\left(\frac{m}{\theta}\right)\right| \geq \frac{n + \sum_{p=0}^{n-1} |q|^p |b_p(\frac{m}{\theta} - n)| |\gamma(\frac{m}{\theta} - n + p)|}{|q|^n |b_n(\frac{m}{\theta} - n)|}$$

with $n = [m\theta^{-1}] - 1$, $m = 1, 2, \ldots$, and

$$b_p(x) = \prod_{j=0}^{p-1} \sin \pi\theta(x + j), \quad b_0(x) = 1.$$

If φ is a continuous solution of (5.2.2), then

$$\varphi\left(\frac{m}{\theta} - n\right) = \sum_{p=0}^{n} q^p b_p \left(\frac{m}{\theta} - n\right) \gamma\left(\frac{m}{\theta} - n + p\right)$$

and hence

$$\left|\varphi\left(\frac{m}{\theta} - n\right)\right| \geq n.$$

Since $m\theta^{-1} - n \in [0,1]$, the latter inequality contradicts the boundedness of φ on $[0,1]$ which means that L is not a surjective operator in $C^0(\mathbb{R},\mathbb{C})$.

Now we use Theorem 4.3 to localize the set $S(L)$.

Theorem 5.9. *Let for some integers $q_1,\dots,q_s$, $1 \leq q_1,\dots,q_s \leq n$, the mappings $F_{q_1},\dots, F_{q_s}$ be C^k-diffeomorphisms and*

$$a_{q_1}(x) \neq 0,\dots,a_{q_s}(x) \neq 0, \quad x \in \mathbb{R}.$$

Let also intervals $V_1, \dots, V_s$ be absorbers for the families $\{ F_{q_1}^{-1} F_j \}_{j\neq q_1}, \dots,$ $\{ F_{q_s}^{-1} F_j \}_{j\neq q_s}$, respectively. Then $S(L) \subset V$ where

$$V = \bigcap_{k=1}^{s} V_k.$$

Proof. We need to prove that if $x \notin V$, then $x \notin \operatorname{supp} l$, $l \in \operatorname{Coker} L$. Assume, on the contrary, that $x \in \operatorname{supp} l$ for a functional $l \in \operatorname{Coker} L$. Let U be a neighborhood of x such that $\overline{U} \cap V = \emptyset$. Then there exists a C^k-function γ vanishing outside U and such that $l(\gamma) \neq 0$. Hence $\gamma \notin \operatorname{Im} L$.

On the other hand, conditions of Theorem 5.9 mean that equation (5.0.1) is invertible with respect to the variables corresponding to $F_{q_1},\dots, F_{q_s}$. According to Theorem 4.3 the local solution $\varphi = 0$ of (5.0.1) at V may be continued as its global solution on $\mathbb{R}$. It means that $\gamma \in \operatorname{Im} L$ which results in a contradiction. □

5.3 Equations with constant coefficients

Here we consider equations

$$L\varphi(x) \equiv \sum_{j=1}^{n} a_j \varphi(F_j(x)) = \gamma(x), \quad a_j \in \mathbb{C},\ a_j \neq 0.$$

A specific character of constant coefficients allows us to investigate it if the estimates (4.2.5) hold starting from $p = 1$.

Theorem 5.10. *Let for some indices* $q_1, \ldots, q_s$ *intervals* $V_1, \ldots, V_s$ *be absorbers for* $\{F_{q_1}^{-1}F_j\}_{j\neq q_1}, \ldots, \{F_{q_s}^{-1}F_j\}_{j\neq q_s}$ *such that*

$$V = \bigcap_{j=1}^{s} V_j$$

is a closed finite interval and let for some q, $1 \leq q \leq n$, *the estimates*

$$\sum_{j\neq q} \left|\frac{a_j}{a_q}\right| |(F_q^{-1}F_j)'(x)|^p \leq \rho < 1, \quad p = 1, \ldots, k, \; x \in V, \tag{5.3.1}$$

hold. Then the operator L *is semi-Fredholm in* $C^k(\mathbb{R}, \mathbb{C})$ *with* dim Coker $L \leq 1$.

Proof. Consider the equation

$$\psi(x) = \sum_{j\neq q} b_j \psi(H_j(x)) + \gamma(x)$$

where

$$b_j = -\frac{a_j}{a_q}, \quad H_j(x) = F_q^{-1}F_j(x).$$

Differentiating we obtain a new equation

$$\psi'(x) = \sum_{j\neq q} b_j \psi'(H_j(x))H_j'(x) + \gamma'(x)$$

with $\gamma' \in C^{k-1}$. Substitution $\psi'(x) = h(x)$ yields an equation which, according to Corollary 4.3, has a global C^{k-1}-solution $h(x)$.

Now let us fix a point $z \in \mathbb{R}$ and set

$$\psi(x) = \int_z^x h(t)dt.$$

This function belongs to C^k and the difference

$$\psi(x) - \sum_{j\neq t} b_j \psi(H_j(x)) - \gamma(x)$$

is constant. If $1 \in \operatorname{Im} L$, then L is surjective. Otherwise we have the direct decomposition

$$C^k(\mathbb{R}, \mathbb{C}) = \operatorname{Im} L + \mathcal{E}$$

where $\mathcal{E}$ is the space of constant functions. In this case L is not surjective, but $\operatorname{Im} L$ is closed and dim Coker $L = 1$. □

Note that if

$$d = \sum_{j=1}^{n} a_j \neq 0,$$

then the constant function $\varphi(x) = d^{-1}$ is a solution of the equation $L\varphi(x) = 1$. Hence if, in addition to (5.3.1), the condition $d \neq 0$ holds, then L is surjective in $C^k(\mathbb{R}, \mathbb{C})$.

Example 5.9. Let C^k-mappings $F_j : \mathbb{R} \to \mathbb{R}$ be such that

$$|F_j'(x)| \leq \sigma < 1, \quad x \in \mathbb{R}, \quad j = 1, \ldots, n.$$

Then every mapping F_j has a unique fixed point x_j, this point is attractive, every closed interval I with $x_j \in \operatorname{Int} I$ is an absorber for F_j and the interval

$$V = [\min_j x_j, \max_j x_j]$$

is the intersection of all absorbers for the family $\{F_j\}$. Hence if

$$\sum_{j=1}^{n} |a_j| < \frac{1}{\sigma},$$

then according to Theorem 5.10 the operator

$$L\varphi(x) = \varphi(x) - \sum_{j=1}^{n} a_j \varphi(F_j(x))$$

is normally solvable in $C^k(\mathbb{R}, \mathbb{C})$, $k \geq 1$. If, in addition, $\sum a_j \neq 1$, then it is surjective. Moreover, in this case L is bijective. Indeed, let $L\varphi_0(x) = 0$ for a C^k-function φ_0, $k \geq 1$. Then

$$|\varphi_0'(x)| \leq \sum_{j=1}^{n} |a_j||\varphi_0'(F_j(x))||F_j'(x)| \leq \sigma \sum_{j=1}^{n} |a_j| \max_K |\varphi_0'(x)|$$

for every compact interval $K \supset V$. Hence $\varphi_0'(x) = 0$, $x \in K$, $\varphi_0(x) = c =$const, and since $\sum a_j \neq 1$, we obtain $c = 0$.

In the case $\sum_j a_j = 1$ we have

$$\dim \operatorname{Ker} L = \dim \operatorname{Coker} L = 1.$$

Example 5.10. Let

$$L\varphi(x) = \varphi(x) - \frac{1}{2}\left[\varphi\left(\frac{x}{2}\right) + \varphi\left(\frac{x}{2} + \frac{1}{2}\right)\right].$$

According to Theorem 5.10 the operator L is normally solvable in $C^k(\mathbb{R}, \mathbb{C})$, $k \geq 1$, and $\dim \operatorname{Ker} L = \dim \operatorname{Coker} L = 1$. More exactly, if $\gamma \in \operatorname{Im} L$ then

$$l(\gamma) \equiv \int_0^1 \gamma(t)dt = 0$$

and the functional l spans the subspace Coker L. If a C^k-function γ satisfies the latter relation, then $\gamma \in \operatorname{Im} L$ and every C^k-solution may be represented in the form

$$\varphi(x) = c + \sum_{s=0}^{\infty} T^s \gamma(x)$$

where

$$T\varphi(x) = \frac{1}{2}\varphi\left(\frac{x}{2}\right) + \frac{1}{2}\varphi\left(\frac{x}{2} + \frac{1}{2}\right)$$

and c is a constant.

Let us show that L is not normally solvable in $C^0(\mathbb{R}, \mathbb{C})$. Indeed, if $L\varphi = \gamma$, then

$$\varphi(x) = T^n\varphi(x) + \sum_{s=0}^{n-1} T^s\gamma(x). \tag{5.3.2}$$

Since

$$T^n\varphi(0) = \sum_{j=0}^{2^n-1} \frac{1}{2^n}\varphi\left(\frac{j}{2^n}\right)$$

is an integral sum for φ on the segment [0,1], it tends to $l(\varphi)$ for every continuous function φ and if $x = 0$, then the sum in (5.3.2) converges to $\varphi(0) - l(\varphi)$. On the other hand, there exists a continuous function γ such that $l(\gamma) = 0$ and the series $\sum_{s=0}^{\infty} T^s\gamma(0)$ diverges. Such a function belongs to $\overline{\operatorname{Im} L}$ but not to $\operatorname{Im} L$.

To prove that a function γ exists, let us consider the sequence of continuous linear functionals

$$l_n(\gamma) = \sum_{s=0}^{n-1} T^s\gamma(0)$$

on the subspace of functions $\gamma \in C^0[0,1]$ with $l(\gamma) = 0$. If there exists $\lim l_n(\gamma)$ for every such γ as $n \to \infty$, then by the Banach–Steinhaus Theorem

$$|l_n(\gamma)| \leq c \max |\gamma(x)| \tag{5.3.3}$$

with a constant c not depending either on n or γ. On the other side, for every $n = 1, 2, \ldots$ there exists a function γ_n satisfying

$$\gamma_n\left(\frac{j}{2^s}\right) = 1, \quad j = 0, 1, \ldots, 2^s - 1, \quad s = 1, \ldots, n,$$

and

$$\max_{x\in[0,1]} |\gamma_n(x)| \le 1, \qquad \int_0^1 \gamma_n(x)\,dx = 0.$$

For such a function we have

$$l_n(\gamma_n) = n \max |\gamma(x)|$$

contradicting (5.3.3).

The operators L in Examples 5.9 and 5.10 are Fredholm in $C^k(\mathbb{R},\mathbb{C})$, $k \ge 1$. The following example shows that under conditions of Theorem 5.10 the subspace Ker L may be infinite-dimensional and hence, generally speaking, L is not Fredholm.

Example 5.11. Let us consider an operator

$$L\varphi(x) = a_1(x)\varphi(x) + a_2(x)\varphi(\lambda x) + a_3(x)\varphi(x+1) + a_4(x)\varphi(\lambda x + 1)$$

with nowhere vanishing C^k-coefficients $a_j(x), j = 1,2,3,4$, and $\lambda \in (0,1)$. Here $F_1(x) = x$, $F_2(x) = \lambda x$, $F_3(x) = x+1$, $F_4(x) = \lambda x + 1$, and (cf., Example 4.8, Section 4.3) $z_+ = z_- = 0$. Thus the segment $V = [z_-, z_+]$ is reduced to the point $z = 0$. If $a_1(x) \equiv 1$, $a_2(x) \equiv -1$, $a_3(x) \equiv 2^{-1}$, $a_4(x) \equiv -2^{-1}$ and $\lambda \in (0, 3^{-1})$, then the operator L satisfies the conditions of Theorem 5.10 with $q = 1$. The space Coker L contains the functional $l(\varphi) = \varphi(0)$ and according to Theorem 5.10 this space is spanned by l. Therefore if $\gamma \in C^k(\mathbb{R},\mathbb{C}), k \ge 1$, and $\gamma(0) = 0$, then $\gamma \in$ Im L.

In order to show that dim Ker $L = \infty$ we note that for arbitrary numbers $a < 0$ and $b > 0$ the interval $V_1 = [a, +\infty)$ is an absorber for the family

$$F_1^{-1}F_2(x) = \lambda x,\ F_1^{-1}F_3(x) = x+1,\ F_1^{-1}F_4(x) = \lambda x + 1,$$

while the interval $V_2 = (-\infty, b]$ is an absorber for the family

$$F_3^{-1}F_1(x) = x - 1,\ F_3^{-1}F_2(x) = \lambda x - 1,\ F_3^{-1}F_4(x) = \lambda x.$$

Every function $\varphi \in C^k(\mathbb{R},\mathbb{C})$ vanishing in a neighborhood of the set

$$W = F_1([a,b]) \cup F_2([a,b]) \cup F_3([a,b]) \cup F_4([a,b])$$

is a local C^k-solution of the homogeneous equation $L\varphi = 0$ at the set $V_1 \cap V_2 = [a,b]$. By Theorem 4.3 there exists a global C^k–solution φ which coincides with φ_0 on the interval

$$F_1(V_1) \cap F_3(V_2) = [a, b+1].$$

If $b < a+1$, then the set $[a, b+1] \setminus W$ is not empty and φ_0 may be an arbitrary function on any closed interval inside this set. Therefore there exists a C^k-solution of the homogeneous equation $L\varphi = 0$ coinciding with φ_0 on such an interval and as a result dim Ker $L = \infty$.

5.4 Equation with affine transformations of argument

In this section we consider equation (5.0.1) with

$$F_i(x) = \lambda_i x + \mu_i, \quad i = 1 \ldots n,$$

in spaces $C^k(\mathbb{R}, \mathbb{C})$. The simplest example

$$\varphi(\lambda x + \mu) - a(x)\varphi(x) = \gamma(x)$$

was investigated in detail in the previous sections. If $\lambda = 1$, and $a(x) \neq 0$, $x \in \mathbb{R}$, the equation is equivalent to

$$\psi(x + 1) = \psi(x).$$

If $\lambda \neq 1$, then F has the unique fixed point

$$x_0 = \frac{\mu}{1 - \lambda}$$

which is hyperbolic if $\lambda \neq -1$. This case is completely covered by Theorem 5.7. In the case $\lambda = -1$ the mapping $F(x) = -x + \mu$ is 2-periodic and the equation takes on the form

$$b(x)\varphi(x) = \gamma(F(x)) + a(F(x))\gamma(x)$$

with

$$b(x) = 1 - a(x)a(F(x)).$$

If $b(x) \neq 0$, $x \in \mathbb{R}$, then the function

$$\varphi(x) = \frac{\gamma(F(x)) + a(F(x))\gamma(x)}{b(x)}$$

is its unique solution. Otherwise additional necessary conditions arise. For instance, if $b(x) \equiv 0$, $x \in \mathbb{R}$, then $\gamma \in \operatorname{Im} L$ if and only if

$$\gamma(F(x)) + a(F(x))\gamma(x) = 0, \qquad x \in \mathbb{R}.$$

In the particular case $a(x) \equiv 1$, $x \in \mathbb{R}$, this means that γ must be odd with respect to the point $x_0 = 2^{-1}\mu$, i.e.,

$$\gamma(-x + \mu) = -\gamma(x),$$

while for $a(x) \equiv -1$, we have

$$\gamma(-x + \mu) = \gamma(x)$$

and γ is even with respect to the same point. If the function $b \not\equiv 0$ vanishes somewhere, then the normal solvability may fail (see Example 5.1).

Let us now consider a general operator

$$L\varphi(x) = \sum_{j=1}^{n} a_j(x)\varphi(\lambda_j x + \mu_j), \qquad a_j \in C^k(\mathbb{R}, \mathbb{C}). \tag{5.4.1}$$

Assume that

$$\lambda_1 = \cdots = \lambda_s > \lambda_j > 0, \quad j \geq s + 1, \tag{5.4.2}$$

for some s, $1 \leq s \leq n$, and

$$\mu_1 < \cdots < \mu_s.$$

If

$$a_1(x) \neq 0, \;\; a_s(x) \neq 0, \quad x \in \mathbb{R},$$

then it follows from part **3** of Section 4.3 and Theorem 5.9 that the interval $[z_-, z_+]$, where

$$z_- = \min_{j>s} \frac{\mu_j - \mu_1}{\lambda - \lambda_j}, \quad z_+ = \max_{j>s} \frac{\mu_j - \mu_s}{\lambda - \lambda_j}, \quad \lambda = \lambda_1 = \cdots = \lambda_s,$$

contains all obstacles to the solvability. In other words, the following implication is true:

$$l \in \operatorname{Coker} L \implies \operatorname{supp} l \subset [z_-, z_+]. \tag{5.4.3}$$

In particular, if $z_+ < z_-$, then the latter interval is empty and $\operatorname{Coker} L = \{0\}$.

Theorem 5.11. *If $z_+ < z_-$, then the operator (5.4.1) is surjective in $C^k(\mathbb{R}, C)$ and dim $\operatorname{Ker} L = \infty$.*

Proof. The surjectivity follows from Corollary 4.1. To prove the second statement let us choose numbers a and b such that $z_+ < b < a < z_-$ and

$$\lambda a + \mu_1 < \lambda b + \mu_s.$$

Then the interval $V_1 = [a, +\infty)$ is an absorber for the family $\{F_1^{-1}F_j\}_{j\neq 1}$ and the interval $V_2 = (-\infty, b]$ is an absorber for the family $\{F_s^{-1}F_j\}_{j\neq s}$. The intersection $V_1 \bigcap V_2$ is empty, since $b < a$. However the intersection $F_1(V_1) \bigcap F_s(V_2)$ contains the non-empty interval $[\lambda a + \mu_1, \lambda b + \mu_s]$. According to Theorem 4.3 for every C^k-function φ_0 there exists a C^k-solution of the homogeneous equation $L\varphi = 0$, coinciding with φ_0 on the interval $F_1(V_1) \bigcap F_s(V_2)$. Hence $\dim \operatorname{Ker} L = \infty$. □

Example 5.12. If

$$\lambda = \lambda_1 = \cdots = \lambda_{n-1} > \lambda_n > 0,$$

then

$$z_- = \frac{\mu_n - \mu_1}{\lambda - \lambda_n} > z_+ = \frac{\mu_n - \mu_s}{\lambda - \lambda_n}$$

and therefore in this case L is a surjective operator with $\dim \operatorname{Ker} L = \infty$.

We will apply now the previous results to a problem of compactly supported solutions of the homogeneous equation $L\varphi = 0$. The following proposition shows that information on such solutions may be extracted from (5.4.3).

Theorem 5.12. *Let the numbers* $\lambda_i > 0$ *and* μ_i *be such that*

$$\tilde{\lambda} = \lambda_n = \lambda_{n-1} = \cdots = \lambda_q < \lambda_j, \quad 1 \le j \le q - 1,$$

and

$$\frac{\mu_q}{\lambda_q} < \frac{\mu_{q+1}}{\lambda_{q+1}} < \cdots < \frac{\mu_n}{\lambda_n}.$$

Assume that

$$a_n(x) \neq 0, \quad a_q(x) \neq 0, \quad x \in \mathbb{R}.$$

If a compactly supported continuous solution of the homogeneous equation $L\varphi = 0$ *with operator* (5.4.1) *is equal to zero on the interval* $[z_-^*, z_+^*]$ *where*

$$z_-^* = \min_{j<q} \frac{\lambda_j \mu_q - \mu_j \tilde{\lambda}}{\lambda_j - \tilde{\lambda}}, \quad z_+^* = \max_{j<q} \frac{\lambda_j \mu_n - \mu_j \tilde{\lambda}}{\lambda_j - \tilde{\lambda}},$$

then it is equal to zero identically on $\mathbb{R}$.

In particular, if $z_+^* \le z_-^*$, *then the homogeneous equation has no non-trivial continuous compactly supported solutions.*

Proof. Let us consider the operator

$$L^*\varphi(x) = \sum_{i=1}^{n} a_i^*(x)\varphi\left(\frac{1}{\lambda_i}x - \frac{\mu_i}{\lambda_i}\right)$$

with

$$a_i^*(x) = \frac{1}{\lambda_i} a_i \left(\frac{1}{\lambda_i}x - \frac{\mu_i}{\lambda_i}\right), \quad i = 1, \dots n.$$

If $\varphi_0 \in \operatorname{Ker} L$ is a continuous compactly supported function, then the expression

$$l_0(h) = \int_{-\infty}^{\infty} h(x)\varphi_0(x)dx$$

defines a linear functional $l_0 \in (C^k(\mathbb{R}, \mathbb{C}))^*$. For $h = L^*g$ we have

$$l_0(h) = \int_{-\infty}^{\infty} \varphi_0(x) \sum_{i=1}^{n} \frac{1}{\lambda_i} a_i \left(\frac{1}{\lambda_i}x - \frac{1}{\lambda_i}\mu_i\right) g\left(\frac{1}{\lambda_i}x - \frac{1}{\lambda_i}\mu_i\right) dx$$

$$= \int_{-\infty}^{\infty} g(t) \sum_{i=1}^{n} a_i(t)\varphi_0(\lambda_i t + \mu_i)\, dt = 0.$$

Therefore $l_0 \in \operatorname{Coker} L^*$. The numbers (4.3.7) for the operator L^* are equal to z_-^* and z_+^*. Applying (5.4.3) to the operator L^* we arrive at the relation $\operatorname{supp} l_0 \subset [z_-^*, z_+^*]$. Hence if φ_0 vanishes in the interval $[z_-^*, z_+^*]$, then $\varphi_0 = 0$. □

Example 5.13. Let $\sigma \in (0,1)$ and

$$L\varphi(x) = a_1(x)\varphi(\sigma x) + a_2(x)\varphi(x+1) + a_3(x)\varphi(x-1) + a_4(x)\varphi(x).$$

Then

$$z_-^* = -\frac{\sigma}{1-\sigma}, \quad z_+^* = \frac{\sigma}{1-\sigma}.$$

If φ_0 is a continuous compactly supported solution of the equation $L\varphi = 0$, then

$$\operatorname{supp}\varphi_0 \subset \left[-\frac{\sigma}{1-\sigma}, \frac{\sigma}{1-\sigma}\right].$$

The equation $L\varphi = 0$ with coefficients

$$a_1(x) = 4\sigma, \quad a_2(x) = a_3(x) = -1, \quad a_4(x) = -2$$

is known as *Schilling's equation.* For $\sigma = 2^{-1}$ it has the continuous solution

$$\varphi_0(x) = \max(1 - |x|, 0)$$

with $\operatorname{supp}\varphi_0 = [-1,1]$.

In the remaining part of the section we consider operators (5.4.1) with constant coefficients. Every such operator is well defined in $C^k(\mathbb{R},\mathbb{C})$ for $k = 0,1,\dots$ and $k = \infty$.

Theorem 5.13. *Let L be an operator of the form (5.4.1) with $a_i(x) \equiv a_i$, $1 \le i \le n$, and let*

$$\sum_{i=1}^{n} a_i\lambda_i^p \neq 0, \quad p = 0,1,\dots. \tag{5.4.4}$$

Then $\operatorname{Im} L$ *is dense in* $C^k(\mathbb{R},\mathbb{C})$ *for every* k *and hence* $\operatorname{Coker} L = \{0\}$.

Proof. It is sufficient to check that (5.4.4) implies $P \in \operatorname{Im} L$ for every polynomial $P(x)$. If c is a constant, then the equation $L\varphi(x) = c$ has the constant solution

$$\varphi(x) = \frac{c}{\sum_{i=1}^{n} a_i}, \quad x \in \mathbb{R}.$$

Hence it is sufficient to assume that $P \in \operatorname{Im} L$ for all polynomials P, deg $P \le k-1$, and to show that the polynomial $P(x) = cx^k$ belongs to $\operatorname{Im} L$ as well. To this end we set

$$\varphi(x) = dx^k + \psi(x), \quad d = \frac{c}{\sum_{i=1}^{n} a_i\lambda_i^k},$$

with a polynomial ψ, $\deg \psi \leq k - 1$. Then

$$L\varphi(x) = cx^k + L\psi(x) + Q(x)$$

where

$$Q(x) = d \sum_{i=1}^{n} a_i(\lambda_i x + \mu_i)^k - cx^k$$

is a polynomial of degree $\leq k - 1$. According to the induction assumption, there exists a polynomial ψ of degree $\leq k-1$ such that $L\psi = -Q$. With such a choice of ψ we obtain $L\varphi(x) = cx^k$. □

If $L\varphi(x) = \varphi(x) - \varphi(x+1)$, then (5.4.4) is wrong for every $p = 0, 1, \dots$, but nevertheless L is a surjective operator in $C^k(\mathbb{R}, \mathbb{C})$ for every k. The following theorem shows that if $s = 1$ in (5.4.2), then condition (5.4.4) is necessary for L to be bijective in $C^\infty(\mathbb{R}, \mathbb{C})$.

Theorem 5.14. *Let L be an operator of the form* (5.4.1) *with constant coefficients $a_i(x) = a_i \neq 0$ and let $\lambda_i > 0, i = 1, \dots, n$. Then there exists a number k_0 depending on the collection $\{a_i, \lambda_i, \mu_i\}_{i=1}^n$ such that L is a semi-Fredholm operator in every space $C^k(\mathbb{R}, \mathbb{C})$, $k \geq k_0$, with* $\dim \operatorname{Coker} L < \infty$. *If, in addition, $s = 1$ in* (5.4.2) *then L is a Fredholm operator in every space $C^k(\mathbb{R}, \mathbb{C})$, $k \geq k_0$, and condition* (5.4.4) *is necessary and sufficient for L to be invertible in it.*

Proof. It is sufficient to prove all statements for an operator of the form

$$L\varphi(x) = \varphi(x) + \sum_{i=2}^{n} b_i \varphi(\alpha_i x + \beta_i)$$

with

$$\alpha_1 = \alpha_2 = \cdots = \alpha_s = 1 > \alpha_j > 0, \quad j \geq s+1,$$

and

$$0 < \beta_2 < \cdots < \beta_s.$$

We start with the simplest case $s = 1$. In this case

$$z_- = \min_{2 \leq i \leq s} \frac{\beta_i}{1 - \alpha_i} < z_+ = \max_{2 \leq i \leq s} \frac{\beta_i}{1 - \alpha_i}.$$

Let us define k_0 as the smallest non-negative integer k satisfying the inequality

$$\sum_{i=2}^{n} |b_i| \alpha_i^k < 1 \tag{5.4.5}$$

and consider the equation

$$\psi(x) + \sum_{i=2}^{n} b_i \alpha_i^{k_0} \psi(\alpha_i x + \beta_i) = \gamma^{(k_0)}(x).$$

It follows from Corollary 4.3 that the latter equation has a solution $\psi \in C^{k-k_0}(\mathbb{R}, \mathbb{C})$. If we set

$$\varphi(x) = \int_{z_-}^{x} \frac{(x-t)^{(k_0-1)}\psi(t)}{(k_0-1)!}\,dt$$

and $P_\gamma = L\varphi - \gamma$, then we obtain $P_\gamma \in C^k(\mathbb{R}, \mathbb{C})$, $P_\gamma^{(k_0)}(x) \equiv 0$ and hence P_γ belongs to the space $\mathbb{C}_{k_0}[x]$ of all polynomials of degree not exceeding $k_0 - 1$. As a result, for every $\gamma \in C^k(\mathbb{R}, \mathbb{C})$ there exists a polynomial $P_\gamma \in \mathbb{C}_{k_0}[x]$ such that $\gamma + P_\gamma \in \operatorname{Im} L$. In other words,

$$C^k(\mathbb{R}, \mathbb{C}) = \operatorname{Im} L + \mathbb{C}_{k_0}[x].$$

Therefore Im L is a closed subspace and dim Coker $L \leq k_0$. Further, if $\varphi \in C^k$ belongs to Ker L, then $\varphi^{(k_0)} = 0$ and $\varphi \in \mathbb{C}_{k_0}[x]$. Hence dim Ker $L \leq k_0$. At last, since the subspace $\mathbb{C}_{k_0}[x]$ is L-invariant, the operator $L : C^k(\mathbb{R}, \mathbb{C}) \to (\mathbb{R}, \mathbb{C})$ is bijective if and only if the restriction

$$L : \mathbb{C}_{k_0}[x] \to \mathbb{C}_{k_0}[x]$$

is bijective. The latter property is equivalent to (5.4.4).

Now we consider the case $s \geq 2$ with $z_- \leq z_+$. The operator L may be written in the form

$$L\varphi(x) = \varphi(x) + T_1\varphi(x) + T_2\varphi(x)$$

with

$$T_1\varphi(x) = \sum_{i=2}^{s} b_i\varphi(x + \beta_i), \quad T_2\varphi(x) = \sum_{i=s+1}^{n} b_i\varphi_i(\alpha_i x + \beta_i).$$

At first let us prove that if a number k_0 is sufficiently large, then the operator

$$L_{k_0}\varphi(x) = \varphi(x) + T_1\varphi(x) + \sum_{i=s+1}^{n} b_i\alpha_i^{k_0}\varphi(\alpha_i x + \beta_i)$$

is surjective in all spaces $C^p(\mathbb{R}, \mathbb{C})$, $p = 0, 1, \ldots,$. According to Theorem 4.3 it is sufficient to prove that for a given $\gamma \in C^p(\mathbb{R}, \mathbb{C})$ the equation

$$L_{k_0}\varphi(x) = \gamma(x)$$

has a local solution at $[z_-, z_+]$. For the proof we choose a C^∞-function τ equal to 1 in a neighborhood of the interval $[z_-, z_+]$ and vanishing outside its neighborhood and consider the equation

$$\varphi(x) = \tilde{T}_1\varphi(x) + \tilde{T}_2\varphi(x) + \gamma(x) \tag{5.4.6}$$

with

$$\tilde{T}_1\varphi(x) = -\tau(x)T_1\varphi(x)$$

and

$$\tilde{T}_2\varphi(x) = -\sum_{i=s+1}^{n} b_i\alpha_i^{k_0}\varphi(\alpha_i x + \beta_i).$$

Since $\tilde{T}_1^m = 0$ for some integer $m \geq 0$, the operator $I + \tilde{T}_1$ is invertible, $(I + \tilde{T}_1)^{-1} = I + \tilde{T}_1 + \cdots + \tilde{T}_1^{m-1}$ and

$$(I + \tilde{T}_1)^{-1}\psi(x) = \psi(x) + \sum_{i=1}^{N} b_i(x)\psi(x + \delta_i), \qquad x \in \mathbb{R},$$

where $\delta_i > 0$ and coefficients $b_i(x)$ are compactly supported C^∞-functions. Therefore (5.4.6) is equivalent to the equation

$$\varphi(x) = \sum_{i=1}^{n} \alpha_i^{k_0} \sum_{j=1}^{N} b_{ij}(x)\varphi(\alpha_i x + \beta_{ij}) + \tilde{\gamma}(x) \tag{5.4.7}$$

with C^∞-coefficients $b_{ij}(x)$ bounded on $\mathbb{R}$ and not depending on a choice of k_0.

Let us choose k_0 such that

$$\sum_{i=1}^{n}\sum_{j=1}^{N} \alpha_i^{k_0}|b_{ij}(x)| \leq \rho < 1, \quad x \in \mathbb{R}.$$

Then (5.4.7) has a solution $\varphi \in C^p(\mathbb{R}, \mathbb{C})$ which satisfies (5.4.6) as well.

Coming back to the operator L with $s \geq 2$, we choose k_0 such that the operator L_{k_0} is surjective in $C^p(\mathbb{R}, \mathbb{C})$ for all $p = 0, 1, \ldots$. Given $\gamma \in C^k(\mathbb{R}, \mathbb{C})$, $k \geq k_0$, consider the equation

$$L_{k_0}\psi(x) = \gamma^{(k_0)}(x).$$

By virtue of a choice of k_0, this equation has a solution $\psi \in C^{k-k_0}$. The function

$$\varphi(x) = \int_{z_-}^{x} \frac{(x - t)^{(k_0-1)}}{(k_0 - 1)!}\psi(t)dt$$

belongs to $C^k(\mathbb{R}, \mathbb{C})$ and the difference

$$P_\gamma(x) = L\varphi(x) - \gamma(x), \quad x \in \mathbb{R},$$

belongs to the space $\mathbb{C}_{k_0}[x]$ of polynomials of degrees not greater than $k_0 - 1$. Now we can complete the proof in the same way as for $s = 1$. □

Theorems 5.13 and 5.14 yield the following proposition.

Corollary 5.6. *If* $\lambda_i > 0$ *and* (5.4.4) *holds, then the operator* $L : C^k(\mathbb{R}, \mathbb{C}) \to C^k(\mathbb{R}, \mathbb{C})$ *is surjective for* $k \geq k_0$.

If $k < k_0$ and (5.4.4) holds, then according to Theorem 5.13 the manifold Im L is dense in the space $C^k(\mathbb{R}, \mathbb{C})$, but may not coincide with it. The simplest example of such a situation is given by equation (3.2.7) with $\lambda^k = |\mu|$, $\lambda^k \neq \mu, \lambda \in (0,1)$. Here $F_1(x) = x, F_2(x) = \lambda x$ and $x = 0$ is a common fixed point for F_1 and F_2. The following example demonstrates that L may not be surjective even if there are no resonances and $F_j, j = 1, 2, \ldots$ have no common fixed points.

Example 5.14. Let (cf., Example 5.10)

$$L\varphi(x) = \varphi(x) - \frac{1}{2}\left[\varphi\left(\frac{x}{2}\right) - \varphi\left(\frac{x}{2} + \frac{1}{2}\right)\right].$$

Here $k_0 = 1$ and since (5.4.4) holds, according to Corollary 5.6 the operator L is surjective in $C^k(\mathbb{R}, \mathbb{C})$, $k \geq 1$. Let us show that L is not surjective in $C^0(\mathbb{R}, \mathbb{C})$.

Indeed, if $L\varphi = \gamma$, then the representation (5.3.2) is valid with

$$T\varphi(x) = \frac{1}{2}\varphi\left(\frac{x}{2}\right) - \frac{1}{2}\varphi\left(\frac{x}{2} + \frac{1}{2}\right).$$

Substituting $x = 0$ in (5.3.2) we obtain

$$\varphi(0) = T^n\varphi(0) + \sum_{k=0}^{n-1} T^k\gamma(0),$$

where

$$T^k\gamma(0) = \frac{1}{2^k}\sum_{j\in S_k^+}\gamma\left(\frac{j}{2^k}\right) - \frac{1}{2^k}\sum_{j\in S_k^-}\gamma\left(\frac{j}{2^k}\right).$$

Here the sets S_k^+ and S_k^- are such that

$$S_k^- \bigcup S_k^+ = \{0, 1, \ldots, 2^k - 1\}, \quad S_1^+ = \{0\}, \quad S_1^- = \{1\}$$

and

$$S_{k+1}^+ = 2S_k^+ \bigcup (2S_k^- + 1), \qquad S_{k+1}^- = 2S_k^- \bigcup (2S_k^+ + 1).$$

Using these relations it is easy to check inductively that the sets

$$M_n^+ = \left\{\frac{j}{2^k}, j \in S_k^+, 1 \leq k \leq n\right\}, \quad M_n^- = \left\{\frac{j}{2^k}, j \in S_k^-, 1 \leq k \leq n\right\}$$

do not intersect.

Let us assume now that L is a surjective operator in $C^0(\mathbb{R}, \mathbb{C})$. Then the sequence of linear functionals

$$l_n(\gamma) = \sum_{k=0}^{n-1} T^k \gamma(0)$$

satisfies an estimate

$$|l_n(\gamma)| \le c \max_{0 \le x \le 1} |\gamma(x)|.$$

On the other hand, for every $n = 1, 2, \dots$ there exists a function $\gamma_n \in C^0(\mathbb{R}, \mathbb{C})$ satisfying

$$\max_{x \in [0,1]} |\gamma_n(x)| \le 1,\ x \in \mathbb{R};\ \gamma_n(x) = 1,\ x \in M_n^+;\ \gamma_n(x) = -1,\ x \in M_n^-.$$

For such a function we have

$$l_n(\gamma_n) = n \max_{0 \le x \le 1} |\gamma_n(x)|$$

and we arrive at a contradiction with the previous inequality.

Example 5.15. Let

$$L\varphi(x) = 4\varphi\left(\frac{x}{\sigma}\right) - (\varphi(x+1) + \varphi(x-1) + 2\varphi(x)), \quad \sigma \in (0,1).$$

Here $\lambda_1 = \sigma^{-1} > 1 = \lambda_2 = \lambda_3 = \lambda_4 = 1$, $k_0 = 1$ and by Theorem 5.14 we conclude that L is a Fredholm operator in $C^k(\mathbb{R}, \mathbb{C})$, $k \ge 1$, with at most 1-dimensional subspaces Ker L and Coker L. Since Ker L contains all constants, we conclude that dim Ker $L = 1$.

To describe the subspace Coker L, let

$$h(z) = l(e^{ixz}), \quad z \in \mathbb{C},$$

be the Fourier transform of a linear functional $l \in (C^k(\mathbb{R}, \mathbb{C}))^*$. If $l \in$ Coker L, then h satisfies the equation

$$4h\left(\frac{z}{\sigma}\right) = h(z)(e^{iz} + e^{-iz} + 2),$$

or

$$h(z) = \cos^2 \frac{\sigma z}{2}\, h(\sigma z).$$

Its solution

$$h(z) = c \prod_{k=1}^{\infty} \cos^2 \frac{\sigma^k z}{2}, \quad c = h(0),$$

is an entire function of exponential type $\sigma(1-\sigma)^{-1}$, bounded on the real line. According to the Wiener–Paley theorem [55] the inverse Fourier transform ω of the function $h(z)(z-\pi\sigma^{-1})^{-1}$ is a square-integrable function with $\operatorname{supp}\omega = [-\sigma(1-\sigma)^{-1}, \sigma(1-\sigma)^{-1}]$ and

$$l(\gamma) = -\int\limits_{-\sigma(1-\sigma)^{-1}}^{\sigma(1-\sigma)^{-1}} \left(i\gamma'(x) + \frac{\pi}{\sigma}\gamma(x)\right)\omega(x)\,dx.$$

Therefore dim Coker $L = 1$.

5.5 Comments and references

The operator approach to a study of linear equations in Banach spaces of continuous and smooth functions was systematically developed by Antonevich in [2], see also [4, 5, 6]. The main ingredient of his approach is the spectral theory of linear operators. Our methods complement those developed in the cited monographs and are applicable to equations in topological (and not only Banach) spaces of functions like $C^k(\mathbb{R})$ and $C^\infty(M)$.

Example 5.2 is taken from [2].

5.1. Theorem 5.5 may be deduced from results proved in [52] where the spectra of weighted shift operators on compact manifolds were investigated. The gluing method for a classification of spectral points for such operators on a segment was used in [3]. Theorem 5.6 is valid in a space of analytic functions endowed with an appropriate topology [32]. A study of equations with periodic points based on Lemma 5.1 first appeared in [21].

The statement from Section 5.1.5 on the normal solvability of cohomological equations with preperiodic mappings may be inverted: if the cohomological equation is normally solvable in the space of continuous functions on a compact topological space M, then $F : M \to M$ is a preperiodic mapping, see [24].

5.2. A localization of obstacles based on the joint dynamical behavior of transformations involved was proposed in [19].

5.3. A transition to equations with derivatives of unknown function was first suggested in [27].

5.4. Examples 4.8 and 5.11 show that the condition $z_+ < z_-$ in Theorem 5.11 is not necessary either for L to be surjective or for Ker L to be infinite-dimensional. Moreover, the same arguments as in Example 5.11 show that an operator

$$L\varphi(x) = a_1(x)\varphi(x) + a_2(x)\varphi(x+2)$$

$$+a_3(x)\varphi(\lambda x) + a_4(x)\varphi(\lambda x + 1) + a_5\varphi(\lambda x + 2)$$

with non-vanishing coefficients possesses both properties: L is surjective and dim Ker $L = \infty$, while $[z_-, z_+] = \{0\} \neq \emptyset$. The following questions may be of interest here:

i) What conditions on the mappings $F_j(x) = \lambda_j x + \mu_j$ *guarantee the surjectivity of operator* (5.4.1) *with arbitrary non-vanishing coefficients?*

ii) What conditions guarantee the property dim Ker $L = \infty$ *for operator* (5.4.1) *with arbitrary non-vanishing coefficients?*

Theorem 5.14 was proved in [27]. An alternative proof of the statement dim Coker $L < \infty$ uses the Fourier transform $F(z) = l(e^{izx})$ of a functional $l \in$ Coker L. The latter is an entire function satisfying the equation

$$\sum_{j=1}^{n} a_j e^{i\mu_j z} F(\lambda_j z) = 0.$$

If all the numbers λ_j are of the same sign, then the space of solutions to this equation analytic at $z = 0$ is finite-dimensional, implying dim Coker $L < \infty$.

The assumption $\lambda_j > 0$, $j = 2, \ldots, n$, in the case $s = 1$, is redundant. The same arguments show that, if $|\lambda_1| > |\lambda_j|$, $j \geq 2$, then the operator (5.4.1) with constant coefficients $a_i(x)$ is Fredholm in $C^k(\mathbb{R}, \mathbb{C})$, $k \geq k_0$.

If $\lambda_1 > 0 > \lambda_j$, $j = 2, \ldots, n$, and

$$T\varphi(x) = \sum_{i=2}^{n} b_i \varphi(\alpha_i x + \beta_i), \; \alpha_i = \lambda_j \lambda_1^{-1} < 0,$$

then by Theorem 5.14 the operator $I - T^2$ is normally solvable in $C^k(\mathbb{R}, \mathbb{C}), k \geq k_0$. According to Corollary 5.5 the operator $I - T$ is normally solvable as well.

It seems that the following conjecture is true:

For every operator (5.4.1) *with constant coefficients there exists a number* k_0 *such that it is normally solvable in* $C^k(\mathbb{R}, \mathbb{C})$ *with* $k \geq k_0$.

Detailed information on the Schilling equation is contained in [39, 12]. Numerous applications of functional-differential and functional equations with affine transformations of argument (in physics they are called equations with rescaling) are discussed in [38]. Compactly supported solutions of such equations play an important role in the wavelets theory and in subdivision processes [36, 37].

Bibliography

[1] D. Anosov. On additive functional equation related to an ergodic rotation of a circle, Izv. AN SSR, **37**, 6, (1973), 1259–1274.

[2] A. Antonevich. Linear functional equations. Operator approach, Minsk, (1988), 230 pp.

[3] A. Antonevich, and G. Belitskii. The essential spectra of a weighted shift operator in the spaces of continuous and differentiable functions. Spectral and evolutionary problems, Vol. 8 (Sevastopol, 1997), Tavria Publ., Simferopol', (1998), 3–12.

[4] A. Antonevich, and A. Lebedev. Functional differential equations: I. C^*-theory. Longman Scientific&Technical, Pitman Monographs and Surveys in Pure and Applied Mathematics, **70**, (1994), 504 pp.

[5] A. Antonevich, M. Belousov, and A. Lebedev. Functional differential equations: I. C^*-applications. Part 1. Equations with continuous coefficients, Pitman Monographs and Surveys in Pure and Applied Mathematics, **94**, (1998), 384 pp.

[6] A. Antonevich, M. Belousov, and A. Lebedev. Functional differential equations: I. C^*-applications. Part 2. Equations with discontinuous coefficients and boundary value problems, Pitman Monographs and Surveys in Pure and Applied Mathematics, **95**, (1998), 414 pp.

[7] S. Aranson, E. Zhuzhoma, and M. Malkin. On the interrelation between smoothness and topological properties of transformations of the circle (theorems of Danjoy's type), Manuscript n.3052-84, deposited at VINITI, Gor'kov. Gos. University, Gor'kii, (1984).

[8] S. Aranson, G. Belitskii, and E. Zhuzhoma. Introduction to the qualitative theory of dynamical systems on surface, AMS, Providence (1996), 325 pp.

[9] V. Arnold. Supplementary chapters in the theory of ordinary differential equations, Nauka, Moscow (1978), 304 pp.

[10] M. Artin. Algebraic approximation of structures over complete local rings, Inst. Hautes Études Sci., Publ. Math., **36**, (1969), 23–58.

[11] I. Baker. Zusammensetzungen ganzer Funktionen, Mathem. Zeitschr., **69**, (1958), 121-163.

[12] K. Baron, and W. Jarczyk. Recent results on functional equations in a single variable, perspectives and open problems, Aequationes Math., **61**, (2001), 1–48.

[13] G. Belitskii. Conjugacy of local C^∞–mappings, Funct. Anal. and Applic., **6**, 1 (1972), 66–67.

[14] G. Belitskii. Functional equations and conjugacy of local diffeomorphisms, Math. Sbornik, **91**, 4 (1973), 565–579.

[15] G. Belitskii. Invariant normal forms of formal series, Funct. Anal. Appl., **13**, 46–67.

[16] G. Belitskii. Functional invariants of diffeomorphisms of smooth manifolds, Dokl. Akad. Nauk Ukrain. SSR, Ser. A (1985), 11, 6–8.

[17] G. Belitskii. Functional moduli of diffeomorphisms of the circle, Ukrain. Math. Journ. **38**, 3 (1986), 369–370.

[18] G. Belitskii. Smoothness classification of one-dimensional diffeomorphisms, Sib. Math. Journ. **27**, 6 (1986), 21–24.

[19] G. Belitskii, Absorbers: definitions, properties and applications, Discrete Dynamics in Nature and Society, **1** (1998), 307–313.

[20] G. Belitskii. Systems of difference equations in continuous and smooth functions, Functional-Differential Eq., **7** (2000), 65–76.

[21] G. Belitskii, and N. Bykov. Spaces of cohomologies associated with linear functional equations, Ergod. Th., Dynam. Syst., **18**, (1998), 343–356.

[22] G. Belitskii, E. Dyn'kin, and V. Tkachenko, Difference equations in Carleman classes, Entire functions in modern analysis, Proc. Isr. Math. Conf., **15**, (2001), 31–36.

[23] G. Belitskii, and Yu. Lyubich. The Abel equation and total solvability of linear functional equations, Studia Math. **127**, (1998), 81–97.

[24] G. Belitskii, and Yu. Lyubich. On the normal solvability of coholomogical equations on compact topological spaces, Operator Theory: Adv. and Appl, **103**, 1998, 75–87.

[25] G. Belitskii, and Yu. Lyubich. On the normal solvability of cohomological equations on locally-compact topological space, Nonlinear Analysis, Minsk, **2** (1999), 44–51.

[26] G. Belitskii, and Yu. Lyubich. The locally translation structure of flows on the wandering point sets, Ergod. Th., Dynam. Syst. (2000), **20**, 633–639.

[27] G. Belitskii, and V. Nikolaevskii. Linear functional equations on the line, Integr. Eq. Oper. Theory, **21** (1995), 212–223.

[28] G. Belitskii, and V. Nikolaevskii. Extension theorem for linear functional-differential equations on the line, Functional-Differential Eq., **3** (1995), 5–17.

[29] G. Belitskii, and V. Tkachenko. On locally-analytic solvability of multidemensional functional equations in the neighborhood of a non-singular point, Dokl. AN SSSR, **314**, 1, (1990), 15–17.

[30] G. Belitskii, and V. Tkachenko. Analytic solvability of multi-dimensional functional equations in a neighborhood of non-singular point, Theor. Functions, Funct. Anal. and Appl. **59** (1993), 7–21.

[31] G. Belitskii, and V. Tkachenko. Solvability of linear difference equations in smooth and real-analytic vector functions of several variables, Integ. Eq. Oper. Theory, **18** (1994), 123–129.

[32] G. Belitskii, and V. Tkachenko. Functional equations in real-analytic functions, Studia Mathematica, **143**, 2 (2000), 153–174.

[33] A.D. Bruno. Analytical form of differential equations, Trudy Mosc. Math. Obshch., **25**, (1971), 119–262; English translation: Trans. Moscow Math. Soc. **25**, (1971), 131–288.

[34] N. Bykov. Smooth classification of structurally stable vector fields on a circle, Teor. Funkt., Funktion. Anal. i Prilozhen., **52** (1989), 110–113; Engl. transl. in J. Soviet Math. **52**, (1990), 5, 3426–3428.

[35] N. Bykov. Smooth equivalence of vector fields on the circle, Dokl. Akad. Nauk Ukrain. SSR, (1991), 3, 5–7.

[36] I. Daubechies, and J. Lagarias. Two-scale difference equations I and II, SIAM J. Math. Anal. **22** (1991), 1388–1410; SIAM J. Math. Anal. **23** (1992), 1031–1079.

[37] G. Derfel, N. Dyn, and D. Levin. Generalized refinement equation and subdivision processes, J. Approx. Theory **80**, Nr. 2 (1995), 272–297.

[38] G. Derfel. Functional-differential and functional equations with rescaling. Operator Theory: Advances and Applications, **8** (1995), 100–111.

[39] G. Derfel, and R. Schilling. Spatially chaotic configurations and functional equations with rescaling, J. Phys. A, **29**, (1996), 4537–4547.

[40] E. Dyn'kin. Pseudoanalytic extension of smooth functions, AMS Trans., **2**, 115, (1980), 33–58.

[41] J. Ecalle. Théorie des invariants holomorphés, Publications Mathematiques d'Orsay, No. 67–7409, (1974).

[42] R.E. Edwards. Functional analysis, Holt, Rinehart and Winston, New York–London, (1965), 1073 pp.

[43] A. Gel'fond. Calculus of finite differences, Moscow, (1952), 479 pp. English translation: International Monographs on Advanced Mathematics and Physics, Hindustan Publishing Corp., Delhi, (1971), 451 pp.

[44] R. Godement. Topologie algébrique et théorie de fisceaux, Hermann, (1958).

[45] L. Gruman. Solutions of difference equations with nonconstant coefficients, Complex analysis, Lecture Notes in Math., **1094**, Springer-Verlag, (1984).

[46] L. Hörmander. Introduction to complex analysis in several variables, D.Van Nostrand Comp., (1966), 208 pp.

[47] H. Herman. Sur la conjugation differentiable des diffeomorphismes du circle a des rotation, Publ. Math., Paris, **49**, (1979), 5–233.

[48] N. Hitchin. Vector fields on the circle, Mech. Anal. and Geometry: 200 years after Lagrange, Elsevier Science Publish., B.V., (1991), 359–378.

[49] S.-T. Hu. Homotopy theory, Academic Press, N.Y.-London, (1959).

[50] H. Kamowitz, and S. Sheinberg. The spectrum of endomorphisms of Banach algebras, J. Funct. Analysis, **4**, (1969), 268–276.

[51] A. Katok, and B. Hasselblatt. Introduction to the modern theory of dynamical systems, Encyclopedia Math. and Appl., Cambridge Univ. Press (1995), 802 pp.

[52] A. Kitover. Operators in $C^{(1)}$ induced by smooth mappings, Funct. Anal. and its Applic., **16**, (1982), 61–62.

[53] M. Kuchma, Functional equations in a single variable, Warszawa (1968), 385 pp.

[54] M. Kuchma, B. Choczewski, and R. Ger. Iterative Functional Equations, Cambridge University Press, (1990), 552 pp.

[55] B. Levin. Lectures on entire functions, Transl. of Math. Monogr., AMS, **150**, 249 pp.

[56] S. Lojasiewicz. Sur le probléme de la division, Studia Math., **8**, (1959), 87–136.

[57] M. Lyubich. Dynamics of quadratic polynomials. I, II. Acta Math. **178**, (1997), 185–247, 247–297.

[58] M. Lyubich. The quadratic family as a qualitatively solvable model of chaos, Notices Amer. Math. Soc. 47 (2000), 1042–1052.

[59] Yu. Lyubich. Linear functional analysis, Encyclopedia Math.Sci., Springer Verlag, **19**, (1992).

[60] S. Mandelbroit. Quasianalytic classes of functions, ONTI, Moscow, (1937), 108 pp.

[61] J. Mather. Commutators of diffeomorphisms, Commun. Math. Helv., **48**, (1973), 195–233.

[62] B. Muckenhoupt. Some results on analytic iteration and conjugacy, Amer. J. Math., **84**, (1962), 161–169.

[63] J. Morawiec, and J. Walorski. On the existence of smooth solutions of linear functional equations, Integr. Eq. Oper. Theory, **39** (2001), 222–228.

[64] R. Narasimhan. Analysis on real and complex manifolds, North-Holland Publ. Comp, Paris, 1968.

[65] Nonlinear Stokes phenomena. Edited by Yu. S. Il'yashenko, Advances in Soviet Mathematics, **14**, AMS, Providence, RI, (1993), 287 pp.

[66] Z. Nitecki. Differential dynamics, MIT Press, Cambridge, Massachusetts and London, England, (1971).

[67] H. Poincaré, Sur les propriétés des fonctions définies par les équations aux différences partielles, Oeuvres, **1**, Gauthier-Villars, Paris, (1929).

[68] G. Polya, and G. Szegö. Problems and theorems in analysis. I. Series, integral calculus, theory of functions, Springer-Verlag, Berlin-New York, (1978), xix+389 pp.

[69] V. Rohlin, and D. Fuks. A beginning course in topology: geometrical chapters, Nauka Publishing House, Moscow, (1977), 487 pp.

[70] A. Scharkowskii. Coexistence of cycles of the continuous transformation of the real line into itself, Ukrain. Math. Journal, **16**, 1, (1964), 61–71.

[71] B.V. Shabat. Introduction to complex analysis. Part II. Functions of several variables, Nauka, Moscow, 1976. 400 pp.

[72] S. Sternberg. Local contractions and theorem of Poincare, Amer. J.Math., **79**, 4, (1957), 809–204.

[73] F. Takens. Normal forms for certain singularities of vector field, Ann. Inst. Fourier, **23**, 2, (1973), 163–195.

[74] A.E. Taylor, and D.C. Lay. Introduction to Functional Analysis, John Wiley & Sons, N.Y., (1980).

[75] S. Voronin. Analytic classification of conformal mappings $(\mathbb{C}, 0) \to (\mathbb{C}, 0)$ with identity linear part, Funct. Anal. and Appl., **15** (1981), 1–17.

Index

Operator Theory - Advances and Applications

Previously published titles

OT 125: **Gil, J.B. / Grieser, D. / Lesch, M.** (Eds.). Approaches to Singular Analysis (2001) ISBN 3-7643-6518-8

OT 124: **Dijksma, A. / Kaashoek, M.A. / Ran, A.C.M.** (Eds.) Recent Advances in Operator Theory (2001) ISBN 3-7643-6573-0

OT 123: **Alpay, D. / Vinnikov, V.** (Eds.) Operator Theory, System Theory and Related Topics (2001) ISBN 3-7643-6523-4

OT 122: **Bart, H. / Gohberg, I. / Ran, A.C.** (Eds.) Operator Theory and Analysis (2001) ISBN 3-7643-6499-8

OT 121: **Elschner, J. / Gohberg, I. / Silbermann, B.** (Eds.) Problems and Methods in Mathematical Physics (2001) ISBN 3-7643-6477-7

OT 120: **Beltita, D. / Sabac, M.** Lie Algebras of Bounded Operators (2001) ISBN 3-7643-6404-1

OT 119: **Litvinov, W.G.**. Optimization in Elliptic Problems with Applications to Mechanics of Deformable Bodies and Fluid Mechanics (2000) ISBN 3-7643-6199-9

OT 118: **Kaashoek, M. / Langer, H. / Popov, G.** (Eds.). Operator Theory and Related Topics (2000) ISBN 3-7643-6288-X

OT 117: **Adamyan, V.M. / Gohberg, I. / Gorbachuk, M./ Gorbachuk, V.** (Eds.). Differential Operators and Related Topics (2000) ISBN 3-7643-6287-1

OT 116: **Gohberg, I. / Goldberg, S. / Krupnik, N.**. Traces and Determinants of Linear Operators (2000) ISBN 3-7643-6177-8

OT 115: **Bercovici, H. / Foias, C.** (Eds.). Operator Theory and Interpolation (2000) ISBN 3-7643-6229-4

OT 114: **Ramírez de Arellano, E. / Shapiro, M.V. / Tovar, L.M. / Vasilevski, N.L.** (Eds.) Complex Analysis and Related Topics (1999) ISBN 3-7643-6228-6

OT 113: **Havin, V.P. / Nikolski, N.K.** (Eds.) Complex Analysis, Operators, and Related Topics (1999) ISBN 3-7643-6214-6

OT 112: **Maz'ya, V. / Nazarov, S. / Plamenevskij, B.** Asymptotic Theory of Elliptic Boundary Value Problems in Singularly Pertubed Domains, Volume II (2000) ISBN 3-7643-6398-3

OT 111: **Maz'ya, V. / Nazarov, S. / Plamenevskij, B.** Asymptotic Theory of Elliptic Boundary Value Problems in Singularly Pertubed Domains, Volume I (2000)ISBN 3-7643-6397-5

OT 110: **Rossmann, J. / Takác, P. / Wildenhain, G.** (Eds.) The Maz'ya Anniversary Collection,Volume 2 (1999) ISBN 3-7643-6202-2

OT 109: **Rossmann, J. / Takác, P. / Wildenhain, G.** (Eds.) The Maz'ya Anniversary Collection, Volume 1 (1999) ISBN 3-7643-6201-4

OT 108: **Dittrich, J. / Exner, P. / Tater, M.** (Eds.) Mathematical Results in Quantum Mechanics (1999) ISBN 3-7643-6097-6

OT 107: **Sakhnovich, L.A.** Spectral Theory of Canonical Differential Systems (1999) ISBN 3-7643-6057-7

OT 106: **Dijksma, A. / Gohberg, I. / Kaashoek, M.A. / Mennicken, R.** (Eds.), Contributions to Operator Theory in Spaces with an Indefinite Metric. The Heinz Langer Anniversary Volume (1998) ISBN 3-7643-6003-8

OT 105: **Bagdasarov, S.**, Chebyshev Splines and Kolmogorov Inequalities (1998) ISBN 3-7643-5984-6

OT 104: **Bercovici, H. / Foias, C.** (Eds.), Nonselfadjoint Operator Algebras, Operator Theory, and Related Topics. The Carl M. Pearcy Anniversary Volume (1998) ISBN 3-7643-5954-4

OT 103: **Gohberg, I. / Mennicken, R. / Tretter, C.** (Eds.), Recent Progress in Operator Theory. International Workshop on Operator Theory and Applications, IWOTA 95, in Regensburg, July 31-August 4, 1995 (1998) ISBN 3-7643-5891-2

OT 102: **Gohberg, I. / Mennicken, R. / Tretter, C.** (Eds.), Differential and Integral Operators. International Workshop on Operator Theory and Applications, IWOTA 95, in Regensburg, July 31-August 4, 1995 (1998) ISBN 3-7643-5890-4

OT 101: **Eidelman, S.D. / Zhitarashu, N.V.**, Parabolic Boundary Value Problems (1998) ISBN 3-7643-2972-6

OT 100: **Foias, C.I. / Frazho, A.E. / Gohberg, I. / Kaashoek, M.A.** (Eds.), Metric Constrained Interpolation, Commutant Lifting and Systems (1998) ISBN 3-7643-5889-0

OT 99: **Ayerbe Toledano, J.M. / Dominguez Benavides, T. / López Acedo, G.** Measures of Noncompactness in Metric Fixed Point Theory (1997) ISBN 3-7643-5794-0

OT 98: **Gohberg, I. / Lyubich, Y.** (Eds.), New Results in Operator Theory and Its Applications. The Israel M. Glazman Memorial Volume (1997) ISBN 3-7643-5775-4

OT 97: **Gorbachuk, M.L. / Gorbachuk, V.I.**, M.G. Krein's Lectures on Entire Operators (1997) ISBN 3-7643-5704-5

OT 96: **Alpay, D. / Dijksma, A. / Rovnyak, J. / Snoo, H. de** (Eds.), Schur Functions, Operator Colligations, and Reproducing Kernel Pontryagin Spaces (1997) ISBN 3-7643-5763-0

OT 95: **Dym, H. / Fritzsche, B. / Katsnelson, V. / Kirstein, B.** (Eds.), Topics in Interpolation Theory (1997) ISBN 3-7643-5723-1

OT 94: **Kadets, M.I. / Kadets, V.M.**, Series in Banach Spaces. Conditional and Unconditional Convergence (1997) ISBN 3-7643-5401-1

OT 93: **Egorov, Y. / Schulze, B.W.**, Pseudo-Differential Operators, Singularities, Applications (1997) ISBN 3-7643-5484-4

OT 92: **Shklyar, A.Y.**, Complete Second Order Linear Differential Equations in Hilbert Spaces (1997) ISBN 3-7643-5377-5

OT 91: **Skubachevskii, A.L.**, Elliptic Functional Differential Equations and Applications (1997) ISBN 3-7643-5404-6

OT 90: **Böttcher, A. / Gohberg, I.** (Eds.), Singular Integral Operators and Related Topics. Joint German-Israeli Workshop, Tel Aviv, March 1-10, 1995 (1996) ISBN 3-7643-5466-6

OT 89: **Egorov, V. / Kondratiev, V.**, On Spectral Theory of Elliptic Operators (1996) ISBN 3-7643-5390-2

OT 88: **van Neerven, J.**, The Asymptotic Behaviour of Semigroups of Linear Operators (1996) ISBN 3-7643-5455-0

OT 87: **Gohberg, I. / Lancaster, P. / Shivakumar, P.N.** (Eds.), Recent Developments in Operator Theory and Its Applications. International Conference in Winnipeg, October 2-6, 1994 (1996) ISBN 3-7643-5413-5

OT 86: **Berezansky, Y.M. / Sheftel, Z.G. / Us, G.F.**, Functional Analysis. Vol. II (1996) ISBN 3-7643-5345-7

OT 85: **Berezansky, Y.M. / Sheftel, Z.G. / Us, G.F.**, Functional Analysis. Vol. I (1996) ISBN 3-7643-5344-9

OT 83: **Antonevich, A.**, Linear Functional Equations. Operator Approach (1996) ISBN 3-7643-2931-9

OT 82: **Constantinescu, T.**, Schur Parameters, Factorization and Dilation Problems (1996) ISBN 3-7643-5285-X

OT 81: **Upmeier, H.**, Toeplitz Operators and Index Theory in Several Complex Variables (1996) ISBN 3-7643-5282-5

OT 80: **Gohberg, I. / Langer, H.** (Eds.), Operator Theory and Boundary Eigenvalue Problems. International Workshop in Vienna, July 27-30, 1993 (1995) ISBN 3-7643-5275-2

OT 79: **Gohberg, I. / Kaashoek, M.A. / van Schagen, F.**, Partially Specified Matrices and Operators. Classification, Completion, Applications (1995) ISBN 3-7643-5259-0

OT 78: **Demuth, M. / Schulze, B.W.** (Eds.), Partial Differential Operators and Mathematical Physics. International Conference in Holzhau (Germany), July 3 -9, 1994 (1995) ISBN 3-7643-5208-6

OT 77: **Lindenstrauss, J. / Milman, V.D.** (Eds.), Geometric Aspects of Functional Analysis. Israel Seminar (GAFA) 1992-94 (1995) ISBN 3-7643-5207-8

OT 76: **Krasnosel'skij, A.M.**, Asymptotics of Nonlinearities and Operator Equations (1995) ISBN 3-7643-5175-6

OT 75: **Huijsmans, C.B. / Kaashoek, M.A. / Luxemburg, W.A. / Pagter, B. de** (Eds.), Operator Theory in Function Spaces and Banach Lattices. The A.C. Zaanen Anniversary Volume (1995) ISBN 3-7643-5146-2

OT 74: **Hagen, R. / Roch, S. / Silbermann, B.**, Spectral Theory of Approximation Methods for Convolution Equations (1994) ISBN 3-7643-5112-8

OT 73: **Feintuch, A. / Gohberg, I.** (Eds.), Nonselfadjoint Operators and Related Topics. Workshop on Operator Theory and Its Applications, Beersheva, February 24-28, 1992 (1994) ISBN 3-7643-5097-0

OT 72: **Gohberg, I. / Sakhnovich, L.A.** (Eds.), Matrix and Operator Valued Functions. The Vladimir Petrovich Potapov Memorial Volume (1994) ISBN 3-7643-5091-1

OT 71: **Basor, E.L. / Gohberg, I.** (Eds.), Toeplitz Operators and Related Topics. The Harold Widom Anniversary Volume. Workshop on Toeplitz and Wiener-Hopf Operators, Santa Cruz, California, September 20-22, 1992 (1994) ISBN 3-7643-5068-7

OT 70: **Demuth, M. / Exner, P. / Neidhardt, H. / Zagrebnov, V.** (Eds.), Mathematical Results in Quantum Mechanics. International Conference in Blossin (Germany), May 17-21, 1993 (1994) ISBN 3-7643-5025-3

OT 69: **Ashyralyev, A. / Sobolevskii, P.E.**, Well-Posedness of Parabolic Difference Equations (1994) ISBN 3-7643-5024-5

OT 68: **Halanay, A. / Ionescu, V.**, Time-Varying Discrete Linear Systems. Input-Output Operators. Riccati Equations. Disturbance Attenuation (1994) ISBN 3-7643-5012-1

For orders originating from all over the world except USA and Canada:

Birkhäuser Verlag AG
c/o Springer GmbH & Co
Haberstrasse 7
D-69126 Heidelberg
Fax: ++49 / 6221 / 345 4229
e-mail:
birkhauser@springer.de

For orders originating in the USA and Canada:

Birkhäuser
333 Meadowland Parkway
USA-Secaucus
NJ 07094-2491
Fax: ++1 201 348 4505
e-mail:
orders@birkhauser.com

Birkhäuser

GPSR Compliance

The European Union's (EU) General Product Safety Regulation (GPSR) is a set of rules that requires consumer products to be safe and our obligations to ensure this.

If you have any concerns about our products, you can contact us on ProductSafety@springernature.com

In case Publisher is established outside the EU, the EU authorized representative is:

Springer Nature Customer Service Center GmbH
Europaplatz 3
69115 Heidelberg, Germany

Batch number: 10577884

Printed by Printforce, the Netherlands